GREAT MYSTERIES

Mysteries of Space

OPPOSING VIEWPOINTS®

GREAT MYSTERIES

Mysteries of Space

OPPOSING VIEWPOINTS®

Look for these and other exciting *Great Mysteries:
Opposing Viewpoints* books:

GREAT MYSTERIES

Mysteries of Space

OPPOSING VIEWPOINTS®

by Richard Michael Rasmussen

Greenhaven Press, Inc.   P.O. Box 289009, San Diego, California 92198-9009

Library of Congress Cataloging-in-Publication Data

Rasmussen, Richard Michael.
  Mysteries of space : opposing viewpoints / by Richard Michael Rasmussen.
    p. cm. — (Great mysteries)
  Includes bibliographical references and index.
  Summary: Explores differing views on such space-related topics as the origin of the universe, the beginnings of life on earth, and the nature of black holes.
  ISBN 1-56510-097-2 (alk. paper)
  1. Cosmology—Juvenile literature. 2. Evolution (Biology)—Juvenile literature. 3. Astronomy—Juvenile literature.
[1. Cosmology. 2. Life—Origin. 3. Astronomy.] I. Title.
II. Series: Great mysteries (Saint Paul, Minn.)
QB983.R37 1994
520—dc20   93-13592
                                                        CIP
                                                        AC

39669- greenhaven 10/94

To my editor, Terry O'Neill
Thank you for your many helpful suggestions
in making this a better book.

Contents

Introduction

This book is written for the curious—those who want to explore the mysteries that are everywhere. To be human is to be constantly surrounded by wonderment. How do birds fly? Are ghosts real? Can animals and people communicate? Was King Arthur a real person or a myth? Why did Amelia Earhart disappear? Did history really happen the way we think it did? Where did the world come from? Where is it going?

Great Mysteries: Opposing Viewpoints books are intended to offer the reader an opportunity to explore some of the many mysteries that both trouble and intrigue us. For the span of each book, we want the reader to feel that he or she is a scientist investigating the extinction of the dinosaurs, an archaeologist searching for clues to the origin of the great Egyptian pyramids, a psychic detective testing the existence of ESP.

One thing all mysteries have in common is that there is no ready answer. Often there are *many* answers but none on which even the majority of authorities agrees. *Great Mysteries: Opposing Viewpoints* books introduce the intriguing views of the experts, allowing the reader to participate in their explorations, their theories, and their disagreements as they try to explain the mysteries of our world.

But most readers won't want to stop here. These *Great Mysteries: Opposing Viewpoints* aim to stimulate the reader's curiosity. Although truth is often impossible to discover, the search is fascinating. It is up to the reader to examine the evidence, to decide whether the answer is there—or to explore further.

"Penetrating so many secrets, we cease to believe in the unknowable. But there it sits nevertheless, calmly licking its chops."

H.L. Mencken, American essayist

Prologue

Mysteries in Space

Mysteries exist in the depths of space, challenging human abilities to solve them. Some of these mysteries are age-old puzzles. Others are based on startling new scientific discoveries. Most of them deal with basic, unanswered questions about the nature, origin, and ultimate fate of the planets, the stars, and life itself. The search for answers represents one of the great adventures of modern times.

Searching the Cosmos

For the first time in history, humans have the technology to unravel many of these great mysteries. New telescopes and satellites are helping make discoveries at a rapid rate. But every time scientists seem to find an answer, a hundred new questions arise. The universe is so large that humans may never fit all its pieces together. Still, the search goes on, as humans seek to understand the wonders of the cosmos and humanity's place in it.

One

How Did the Universe Begin?

(Opposite page) When gazing into the vast abyss of space, one cannot help but wonder "Where did the universe come from?"

For most of recorded history, humans thought the universe consisted only of the earth and the stars overhead. To ancient peoples, the sky was a solid canopy, enclosing a flat earth on all sides. The stars were mere pinpoints of light, perhaps the perfect light of God shining through the canopy. The sun was a lamp to light the earth.

Today, scientists know that the universe is a much more complicated place. It includes everything, from the small (the atoms and molecules of the human body), to the large (the earth, the sun, the stars, and distant galaxies). The universe is large beyond comprehension. It extends billions of light-years in all directions. (A light-year is the distance light, traveling at 86 thousand miles per second, travels in a year. Thus, a light-year equals nearly 6 trillion miles.)

A person standing on earth and looking at the nighttime sky gazes only into our particular "neighborhood" of the universe—the stars of our own Milky Way galaxy. The sight of millions of distant lights inspires people to ask the same questions the ancients asked: How did all this come to be? Where did the universe come from? Was it always the same as it is now? Where is it going? For centuries scien-

tists and philosophers have tried to solve these riddles. But it is only recently that significant progress has been made in developing scientific accounts.

At the beginning of the twentieth century, most cosmologists—those who study the universe as a whole—believed that the universe existed in a steady, relatively unchanging state. All the parts of the universe appeared to work in perfect harmony with one another, like the parts of a clock. The planets circled the sun in nearly perfect circles. The stars remained steadily in place. This clockwork universe seemed to reflect the perfect design of God. For the most part, it reflected the religious and philosophical thought of previous centuries. Little observable evidence challenged this view.

However, at about this time—the early twentieth century—astronomers were beginning to study light from the stars. They recorded light with special instruments and broke it down into various spectra, or wavelengths. A light wavelength is somewhat like a wave of water at the beach. Each wave has a crest, or high point, and a trough, or valley, which follows

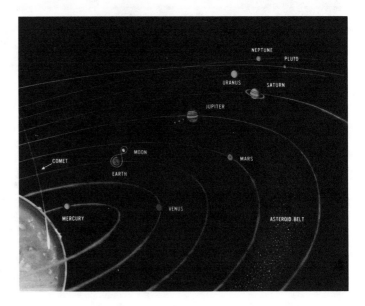

Less than a century ago cosmologists believed that the universe existed in a steady, relatively unchanging state.

the crest. The distance between the crest and the trough represents the length of the wave. Scientists call the range of light wavelengths the electromagnetic spectrum. It extends from short wavelengths (blue) to long wavelengths (red).

Scientists found that different elements in stars produced different wavelengths. The length of the light waves they observed coming from distant stars could tell astronomers what elements produced the light—hydrogen or helium, for example. Scientists now had a way to determine the chemical makeup of distant stars. This opened the door to a rapid series of discoveries—and a new picture of the universe.

The Red Shift

In 1929 astronomer Edwin Hubble made an important discovery. As he recorded light emitted, or put out, from the atoms of distant galaxies, he noticed something peculiar. Instead of matching up with known wavelengths, the light from these galaxies was slightly off. That is, the light waves shifted toward the red end of the spectrum. Hubble also noticed that the farther away the galaxy, the larger the shift toward the red. He attributed this strange phenomenon to something like the Doppler effect.

The Doppler effect was first discovered in relation to sound waves. You experience the Doppler effect when a car speeds by. As the car approaches, its sound seemingly increases in pitch. As the car speeds away, the sound seemingly decreases in pitch. This happens because sound waves have a regular pulse, or frequency. When the car approaches, your ears hear more pulses per second, or a higher pitch. When the car departs, your ears hear fewer pulses per second, or a lower pitch.

Hubble discovered that light waves act similarly to sound waves. In other words, light coming from a star or galaxy that is speeding away from earth will

In 1929 astronomer Edwin Hubble discovered that the galaxies were moving away from earth. His discovery suggested that the entire universe was expanding.

register at a lower than expected frequency. It shifts toward the red end of the spectrum. The red shift of distant galaxies that Hubble observed told him that the galaxies were moving away from earth. This suggested to him and other scientists that the entire universe was expanding.

An Expanding Universe

Imagine that the universe is a piece of bread dough with raisins scattered throughout. Also imagine that each raisin is a galaxy. The dough represents the space between galaxies. As the dough rises, the raisins began to move away from one another. To someone standing on a raisin, a neighboring raisin would seem to be moving away. This is because the dough between the two raisins is constantly expanding. A neighboring raisin twice as far away would appear to be moving away twice as fast, because there is twice as much dough in between. The further away the raisin, the faster it would seem to move away, or recede. It is this behavior that Hubble observed in space.

The model of an expanding universe established by Hubble's observations has three key elements. First, the earth plays no special role in the expansion. Using the dough analogy, an observer would see the same thing no matter what raisin he or she stood on. It would seem that his or her raisin was motionless and all the others would seem to be moving away.

Second, the raisins do not move *through* the dough. Instead, they are carried along *with* the dough as it expands. Thus the galaxies do not move apart from one another through space, but are carried along as space itself expands.

Third, the raisins themselves do not expand. It is only the space between them that grows. Thus the earth and the Milky Way galaxy and other objects in space are not expanding—just the space between them.

The Big Bang

Hubble's theory became known as the Hubble expansion. It gave scientists evidence that the universe did not exist in a steady, unchanging state. It suggested that some event in the past set the expansion in motion. This event, some scientists came to believe, was an explosion, known as the big bang. Today, supporters of the big bang believe it occurred about 15 to 20 billion years ago.

Prior to the explosion, theorists say, all the matter that is now in the universe—the earth, the sun, the solar system, the Milky Way galaxy, and the hundred billion or so other galaxies—existed in a space smaller than that of a needle point. This matter consisted mostly of unimaginably hot, intense radiation. The explosion flung the matter outward in all directions. Elementary particles came into existence about one-millionth of a second after the

"It's impossible that the Big Bang is wrong. Perhaps we'll have to make it more complicated to cover the observations, but it's hard to think of what observations could ever refute the theory itself."

Joseph Silk, astrophysicist, University of California, Berkeley

"If one starts with the present and attempts to go backward in time, there is no reason to assume that there ever was a Big Bang or that the universe had any beginning."

Eric J. Lerner, physicist and independent researcher

explosion. These particles were protons, neutrons, and electrons, the basic building blocks of atoms. Within a mere minute and a half, temperatures dropped enough for atoms of deuterium (a form of hydrogen) to form.

The radiation rapidly expanded outward in what scientists call a primordial fireball. This early period of the universe is known as the radiation or energy era. Heat and energy alone dominated the universe.

After several hundred years, temperatures in the expanding fireball cooled enough for hydrogen to form, and then helium. From the atoms of these and a few other elements, matter, the material that things are made of, was formed. Thus the radiation era ended, and the matter era began. This matter era is still at work today.

Scientists say the ultraviolent big bang created considerable turbulence. Because of this, matter was not distributed evenly as it expanded outward. Thus, some regions were considerably denser, containing

A halo of particles called neutrinos surrounds the Milky Way galaxy. Big bang theorists believe that our galaxy is surrounded by billions of these particles, which were produced during the first few moments of creation.

The Big Bang Theory

 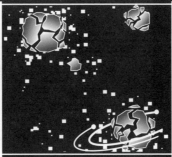 

1. All the matter in the universe was once part of a single, hot, dense object called the primordial egg. This object exploded, sending hot gases and debris outward in all directions.

2. Gradually, the gravity of the larger masses attracted smaller masses to them, forming stars and planets.

3. As the stars continue to hurtle through space, some planets are moving at just the right speed to be drawn into orbits around them. Just as planets are captured in orbit around stars, smaller bodies, or moons, are captured in orbit around the planets. Similarly, entire solar systems are captured in orbits around galaxies, like the Milky Way. And entire galaxies revolve around still larger galaxies.

large amounts of matter. Other areas were less dense, with relatively small amounts of matter. The dense areas slowly started to attract additional matter. About 1 billion years following the explosion, the dense areas formed into galaxies—large, revolving collections of matter. Eventually, the gravity of individual galaxies began to attract other galaxies. The galaxies thus tended to form into large groups, or clusters. Many clusters formed, with a large amount of space between them. In turn, the clusters formed into even larger superclusters. In another 3 billion years, the first stars began forming in individual galaxies.

A major part of the expansion following the big bang probably included the formation of dark matter, material invisible to the naked eye. Modern supporters of the big bang theory are certain this mysterious invisible matter exists. But they do not know exactly what it is made of or why it is not readily detectable. They are certain of its existence only because they have detected its gravitational effect upon visible objects.

Many scientists believe dark matter is the "glue" that holds the universe together. During the early formation of the universe, visible matter probably formed around the dark matter. Scientists estimate that dark matter makes up 90 to 99 percent of the universe. Some of this matter may exist in the halos of dust and gas that surround galaxies. Other dark matter may exist in the depths of black holes or in the space between galaxies.

Today's universe, say big bang scientists, is the result of the original explosion and the expansion that is still going on. As a result, the universe features millions of galaxies, galaxy clusters, and superclusters. Enormous gulfs of black, empty space fill the void in between.

The Steady State Theory

After Hubble's discovery of the red shift, the big bang theory quickly dominated scientific thought. But not all scientists accepted it. Some of them developed an alternate idea, a modern version of the old steady state theory. The new version agrees with the big bang idea that the universe is expanding. But even so, say those who support the steady state theory, the universe always *appears* the same. As galaxies move apart, new matter appears between them. From this new matter, new galaxies form to replace those that have receded into infinite distances. It is the birth of new matter, not a big bang, that is causing the expansion. Astronomer Fred Hoyle explains, "Atoms appear one by one. Instead of the whole universe being created in a flash, in a big bang, atoms are created individually and continuously, with the process of creation going hand-in-hand with the expansion of the universe." The

"The media death of the Big Bang was greatly exaggerated. For real cosmologists, these days, the major question is not whether the Big Bang occurred, but rather, how."

Dennis Overbye, physics journalist

"Practically everybody now goes around talking about the big bang as if it were the one and only cosmology, and very little is said about alternatives. I find this total lack of balance very bad."

Geoffrey R. Burbidge, University of California, San Diego

Astronomer Fred Hoyle believes that it is the birth of new matter, not a big bang, that is causing the expansion of the universe.

universe, says Hoyle, goes on forever. It had no beginning, and it will have no end.

Background Radiation

One piece of evidence supporters of both theories argue over is the significance of background radiation in space. Writer Dennis Overbye argues that cosmic background radiation provides support for the big bang, the "ultimate beginning of beginnings." This background radiation is "a faint uniform microwave hiss that permeates the sky." It comes in small but equal amounts from all distant points in the universe. Big bang scientists predicted the presence of such background radiation at mid-century, but no one actually detected it until 1965.

Measurements taken then indicated that the radiation has the characteristics expected of a fading big bang. In a sense, it is a leftover echo from the original explosion. To big bang supporters, this seemed a final confirmation of their theory.

But critics found problems in repeated measurements of the background microwave radiation. The radiation seemed to be smooth in nature. It did not seem to have any ripples or fluctuations in it. Irregularities would be expected of a universe that quickly became clumpy with many large-scale structures. Therefore this finding seemed to cast doubt on microwave radiation being proof of the big bang.

Fred Hoyle is one of the prime supporters of the steady state theory. He says that acceptance of background microwave radiation as evidence for the big bang "is really just a convention, a way of skirting the fact that our knowledge of the universe is still scant and tentative." Hoyle has a steady state explanation for background microwave radiation. This

Background radiation can be analyzed using a microwave sky map like this. Such data has been used as evidence of the big bang theory.

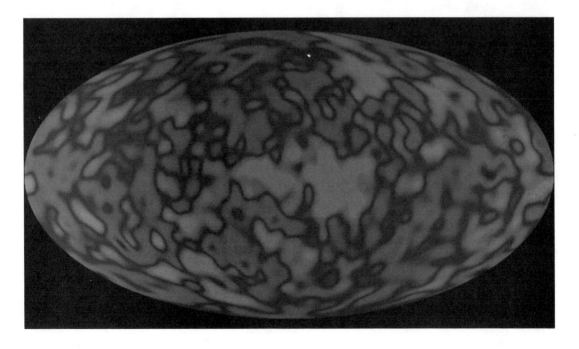

radiation, he says, is merely a transformed version of starlight. The light from stars shines on fine particles floating between the galaxies. Then the starlight is scattered into what we detect as background radiation.

Hoyle finally concludes that neither the big bang nor the steady state theory seems to be quite right about background radiation. The steady state theory predicted background radiation, he said, "but expected it to be in the form of starlight, not in the form of radio waves." The big bang theory predicted a microwave background, but "it was expected to be between ten and a thousand times more powerful than is actually the case."

He adds, "The question now is which of these two inaccuracies is really the less damaging to the theory from which it arises."

A Plasma Universe?

Some scientists were not happy with either of the dominant theories. Some of them came up with alternatives that have influenced scientific thought. For example, independent researcher and big bang critic Eric J. Lerner wrote *The Big Bang Never Happened* in 1990. In this book Lerner is critical of both the big bang and steady state theories. He says that the universe is dominated by electromagnetic energy. The large-scale objects of the universe, such as galaxies, are shaped by the forces of electricity and magnetism. Agreeing with the steady state theorists, Lerner says there never was a big bang. He also agrees that the universe is infinite; it goes on forever. But unlike the steady state theorists, Lerner says that the universe constantly changes and evolves.

The universe, says Lerner, "begins from the known fact that over 99 percent of the matter in the universe is plasma—hot, electrically conducting gases." It is, he says, "a universe crisscrossed by vast electrical currents and powerful magnetic

"For such a simple theory, the big bang theory works amazingly well. It explains the expansion of the universe. It predicts the correct abundances of hydrogen and helium. But perhaps most importantly, it predicts the microwave background radiation."

Richard Talcott, assistant editor, *Astronomy* magazine

"As a general scientific principle, it is undesirable to depend crucially on what is unobservable to explain what is observable, as happens frequently in big bang cosmology."

Halton C. Arp, astrophysicist, 1990

fields, ordered by the cosmic counterpoint of electromagnetism and gravity." In other words, electrical currents, magnetic fields, and gravity influence the shape and behavior of hot gases. Lerner finds support for his theories in the observation of galaxies, quasars, and stars—all objects featuring hot gases.

Big Bang Attacked

Lerner relishes attacking big bang supporters. He says:

> Crucial observations have flatly contradicted the assumptions and predictions of the big bang. Because the big bang supposedly occurred only about 20 billion years ago, nothing in the cosmos can be older than this. Yet in 1986 astronomers discovered that galaxies compose huge agglomerations a billion light years across; such mammoth clusterings of matter must have taken a hundred billion years to form.

Additionally, Lerner charges that an essential component of big bang theory—dark matter—simply does not exist. "The phenomena that the big bang seeks to explain with a mysterious ancient cataclysm, plasma theories attribute to electrical and magnetic processes occurring in the universe today." In other words electromagnetic forces rather than dark matter serve as the glue binding the universe together. These plasma forces, not the gravity and dark matter of the big bang, allow galaxies and other large, clumpy objects to form.

Lerner also gives an explanation for the Hubble red shift. As light travels through space, he says, it loses energy. It is the decline in energy, not the expansion of the universe, that causes the red shift.

Not surprisingly, big bang scientists have contested Lerner's attacks. University of Hawaii physicist Victor J. Stenger says, "Lerner does not make much of a case against [the big bang]. In fact, a great deal of what he discusses in his book, like cosmic

"Support for the theory of the Big Bang comes from observations of a celestial background of relic radiowaves. . . . [This] background radiation is considered by most cosmologists to be the most important confirmation of the Big Bang."

James Trefil, physicist, George Mason University, Fairfax, Virginia

"There is a tremendous bandwagon rolling, which makes anybody who says anything contrary to the big bang highly suspect."

Geoffrey R. Burbidge, University of California, San Diego

Critics of the big bang maintain that galaxies, which are composed of huge clusters a billion light years across, must have taken much more than 20 billion years to form.

plasma phenomena, is perfectly consistent with the big bang. He could have used the same material had he decided to write *The Big Bang Happened!*"

More Assault on the Big Bang

But big bang opponents were not yet ready to quit. These opponents include Geoffrey R. Burbidge of the University of California, San Diego; N.C. Wickramasinghe of the University of Wales College of Cardiff, in Great Britain; German astrophysicist Halton C. Arp; and Fred Hoyle. Together they wrote a paper that shook up the scientific world. Published in the October 30, 1990, issue of *Nature*, the paper, in essence, sought to revise the steady state theory and shoot down the big bang.

"Instead of a static universe we are dealing with a universe that is continually evolving in time. We must think of the universe as process."

James S. Trefil, physicist, George Mason University, Fairfax, Virginia

"In contrast to the Big Bang universe, the plasma universe . . . is formed and controlled by electricity and magnetism, not just gravitation."

Eric J. Lerner, physicist and independent researcher

According to the revised model proposed by these scientists, matter is continually created through a series of "little bangs." New galaxies form at a rate determined by the pace at which the universe expands.

The scientists concluded that quasars—distant radiation sources—are presently much closer than their red shifts indicate. For many years scientists have thought that quasars exist at the far reaches of the universe. They thought quasars were ancient objects, relics of the early universe. As such they were evidence supporting the big bang. Suddenly, the possibility that quasars are not distant objects seemed to cast doubt on the big bang. The scientists questioned that the universe has changed much at all. If it has not changed, then perhaps the steady state theory was the best explanation after all.

Following the paper's publication, newspapers screamed that the big bang had been overthrown. Headlines such as "Astronomers' New Data Jolt Vital Part of Big Bang Theory" and "Big Bang Blown to Pieces" gathered attention worldwide. Such headlines, however, may have been premature.

The COBE Discovery

In April 1992, a research team, including astronomer George Smoot of the University of California, Berkeley, announced an important discovery. The highly sensitive Cosmic Background Explorer (COBE) satellite had provided important new data. It detected small variations in the temperatures of background microwave radiation. (All prior measurements had indicated a smooth background. That is, the temperature remained steady.) Smoot says, "These small variations are the imprints of tiny ripples in the fabric of space-time put there by the primeval explosion. Over billions of years, the smaller of these ripples have grown into galaxies, clusters of galaxies, and the great voids of space." In other words, the ripples are evidence of a clumpy structure from the very beginning, rather than the smooth structure predicted by the steady state theory.

The discovery strongly supports the big bang theory and the existence of dark matter, the invisible fabric that binds the universe together and permits the existence of large-scale structures such as galaxies. University of Pennsylvania cosmologist Paul Steinhardt hailed the COBE discovery. He called it "not only one of the most important discoveries in cosmology, but one of the most important discoveries in science in this century."

Science writer Richard Talcott summarizes: "The discovery is far more than a single piece of the

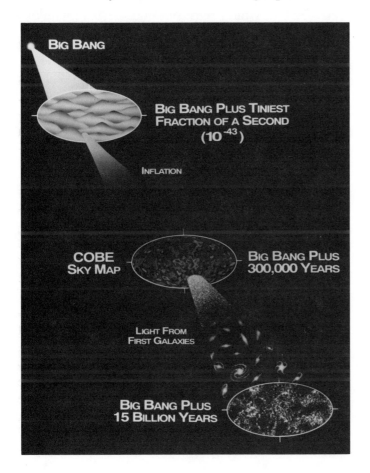

BIG BANG

BIG BANG PLUS TINIEST
FRACTION OF A SECOND
(10^{-43})

INFLATION

COBE
SKY MAP

BIG BANG PLUS
300,000 YEARS

LIGHT FROM
FIRST GALAXIES

BIG BANG PLUS
15 BILLION YEARS

An artist's depiction of the crucial periods in the development of the universe according to the big bang theory, based on COBE's observations of background microwave radiation.

cosmological puzzle. It provides direct support for the theory of [expansion]. . . . And even more importantly, it eliminates a large number of competing models."

What Is the Fate of the Universe?

Regardless of the theory supported, nearly everyone acknowledges that the universe is expanding. One of the greatest mysteries is where it is going and whether it will go on forever.

Most cosmologists believe the universe has two likely fates. One is that it will continue to expand. But as it does so, the rate of expansion will slow down. This universe would gradually become less active. Stars would burn out and die, and galaxies would become thin veils of gas and dust. Eventually, more than 100 billion years from now, the universe essentially would become dead.

The second possibility is that the universe will eventually stop expanding, overcome by its own gravity. It will then begin contracting. Eventually it will compress into an intensely hot, dense nucleus, in a final cataclysmic "big crunch."

Some scientists envision a third possible fate, called the oscillating or pulsating universe. The universe will stop expanding and contract to a big crunch. Then the big crunch will ignite a brand new big bang, and expansion will begin all over, with a new universe. This process of expansion, contraction, and explosion of new universes could go on forever.

Scientists say the possibilities of expansion, contraction, and new expansion are likely only if the universe contains enough matter. If there is enough matter, its gravity will eventually overcome the force of the original explosion, causing expansion to stop, and the universe to begin to contract.

Most scientists believe that visible matter—stars, galaxies, quasars—composes only about 1 to 10 percent of the matter necessary to halt the expan-

"Throughout the universe, the [steady state] theory predicts, new galaxies are forming from atoms that are perpetually being created, and so the telescopic survey would show new galaxies appearing [over time] to replace the old."

Fred Hoyle, British astronomer

"Existence of the cosmic background radiation, together with the spread of galaxies in space, discredits the steady-state hypothesis as a feasible model of the universe."

Eric Chaisson, astrophysicist, Harvard University

Will the universe contract into a final cataclysmic "big crunch," or will it eventually die out? The fate of the planets, and of the entire universe, remains a mystery.

sion. This is where the dark matter many scientists believe exists comes into play. Dark matter could provide enough force to eventually bring about contraction of the universe.

Questions Remain

"The big bang is a hypothesis" (theory), says George B. Field of the Harvard-Smithsonian Center for Astrophysics. "Like any scientific hypothesis, we're trying to verify whether it's true or false by experimental means." He adds, "Precisely because the big bang is the leading model, it should be tested in every possible way in order to find out if it is flawed. It may turn out that you can do a crucial experiment that would disprove it, in which case you would turn to the next best model, whatever it might be."

Edward W. Kolb, of the Fermi National Accelerator Laboratory in Batavia, Illinois, concludes, "There are probably no deeper, more profound questions than the questions asked of cosmology. No matter what the model is, and no matter what you do, it's sure to excite strong feelings in people. No model is going to please everyone."

Two

How Did Life Form on Earth?

While cosmologists have searched the skies for clues to the universe's origin, biologists and other life scientists have searched the earth for clues to life's origin. Life scientists, however, have one advantage over cosmologists. They have a long chain of clues—fossils, the imprints of ancient creatures left in the oldest rocks of the earth. Fossils range from the single-cell and multicelled organisms of ancient earth to complex plants and animals of a prehistoric world. They show a pattern of evolution, or change, from simple to complex, over long periods of time.

The earth is now home to an enormous variety of living things—more than 2 million species. These living things range in size from microbes to massive sequoia trees. Life is found in a multitude of environments, from icy polar waters to blazing deserts. It occurs in every imaginable form and exhibits a wide range of complex behaviors and adaptations. Yet, the fossil record shows that this was not always so. How did all this life come to be?

Two Main Scientific Theories

Modern scientists have developed two main theories for explaining the development of life on earth. Both are influenced by or linked to events

occurring in outer space. The first and most widely accepted explanation is that of chemical evolution. According to this theory, life arose through a sequence of chemical actions during earth's early history. The second theory, called panspermia, says that life came from outer space in the form of cosmic seeds. These seeds settled on earth's surface eons ago and started the process of biological change.

A Chemical Basis for All Life

Most scientists, whether supporters of chemical evolution or panspermia, agree that all life on earth shares common characteristics. These include the ability to reproduce, to metabolize or chemically process molecules and energy into usable forms, to grow, to sense and respond to changes in the immediate surroundings, and to adapt to changes from generation to generation. In all likelihood, scientists say, all life also shares a common beginning.

It is difficult to grasp how life could have developed from what was once nonliving matter. However, this is what the theory of chemical evolution attempts to do.

Chemical Evolution

The story of chemical evolution begins with earth's formation. Most scientists believe the sun, the earth, and the other planets of our solar system were created about 5 billion years ago. They formed from a disc of cosmic dust and interstellar gas swirling in the midst of deep space. This cosmic dust came from the explosions of old stars. The dust contained, among other things, the heavier elements and compounds necessary for planet formation. These include iron and water.

Through the random attraction of gravity, some of the cosmic dust in the disc began gathering into clumps. Most of this material drifted toward the disc's center, where gravitational pressures caused

"Chemical properties [in the early oceans of earth] led to incredibly complex 'behaviors' that mimic those of simple lifeforms."

William K. Hartmann, planetary scientist, University of Arizona

"Biochemical systems are exceedingly complex, so much so that the chance of their being formed through random shufflings of simple organic molecules is exceedingly minute, to a point indeed where it is insensibly different from zero."

Fred Hoyle, British astronomer

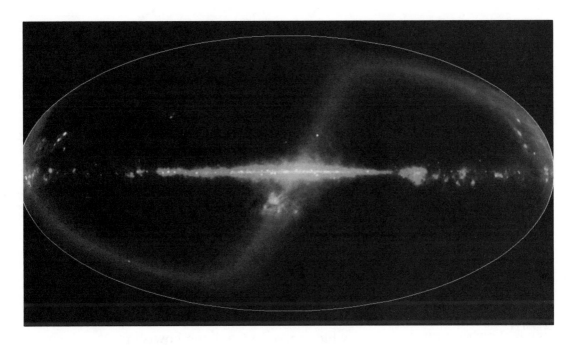

COBE captured this image of the dusty regions of the Milky Way galaxy. Scientists believe that our solar system was formed from a disc of cosmic dust that came from the explosions of old stars.

temperatures to soar. This hot, central part of the disc began to evolve into a proto-sun, or primitive sun. When the proto-sun grew large enough, the increased heat and pressures ignited thermonuclear processes deep in its interior. As this new energy radiated to the surface, the sun began to shine.

Meanwhile the smaller, individual clumps of matter circling the sun continued to sweep up loose material. The clumps grew larger over time. As these clumps thickened and cooled, they formed into the planets of our solar system. One of these planets was the earth. Thus, like the other planets, everything on the earth—every atom, molecule, and chemical element—was born of the material from ancient stars.

Conditions on Early Earth

Scientists agree that the conditions on early earth were very different from those that exist today. The atmosphere was unbreathable by current standards. It probably was composed of large amounts

Scientists believe that the ocean of prehistoric times was a kind of primordial soup—a rich mixture of organic molecules from which life spontaneously arose.

of hydrogen, ammonia, methane, water, and carbon dioxide. The earth itself was mostly covered with water. In places, islands of heavily volcanic, barren land rose above the waves.

Research scientist Robert Hazen finds it remarkable that life evolved at all. He notes, "On a clear winter's night, gazing into the cold depth of the sky, you can face the brute fact that the universe is a cold, hostile, forbidding place." Few places, he says, offer the proper conditions for life to evolve. These conditions must include just the right temperature, pressure, and chemical elements, as well as a source of energy to combine these elements. "The early earth," he says, "provided all those conditions."

Indeed, most scientists agree that all of the essential building blocks of life were present on early earth. These included the oceans and tidepools, which served as "mixing bowls" for the chemicals of life. Temperatures were right, remaining above freezing and below boiling, so that life-forming processes could take place in the water.

And, say supporters of chemical evolution, the six essential elements for life's evolution were in abundant supply. These elements were carbon, hydrogen, nitrogen, oxygen, phosphorus, and sulfur, present in the water. Early atmospheric gases, in addition to carbon dioxide, likely included ammonia, methane, hydrogen, and water vapor. Some of these elements were added to the early carbon dioxide atmosphere by erupting volcanoes. Robert Hazen says, "These gases mixed with the wave-tossed surface layers of the early ocean, which thus contained all the essential elements of life."

The Primordial Soup

About 3.5 billion years ago, scientists say, simple gases in the oceans gradually became energized by two sources. One source was repeated strikes of lightning from the sky. Another was the steady dose of radiation from the sun. These energy sources enabled the gases in the water to combine into more complex carbon-based molecules. For hundreds of millions of years, this layer of molecules covered the upper layer of the oceans. Scientists have labeled this molecule layer primordial [ancient] soup. Probably included in this primordial soup were a variety of sugars and amino acids, later important to the functions of living cells.

Some of these molecules, simply by attracting other atoms and then breaking apart, began to manufacture copies of themselves. This was an early chemical form of reproduction. Once this process took hold, huge concentrations of these larger molecules soon dominated the oceans. These molecules were not genuine living matter. But they could be said to be feeding because of their random absorption of carbon, hydrogen, oxygen, and nitrogen atoms.

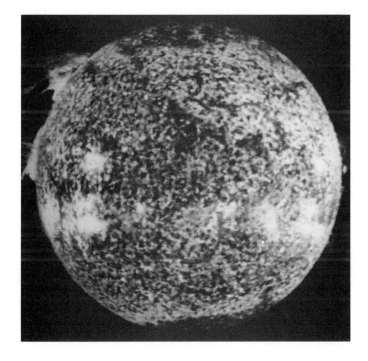

Flares of intense heat erupt from the sun. Scientists believe the sun's radiation provided energy required to transform simple gases in the water into early life-forms.

Eventually, scientists speculate, larger molecules with protective chemical shells and a hollow center appeared. These molecules could pass sea water to their interiors. They could absorb needed atoms and molecules for growth. And they could shield themselves from absorption by larger molecules. These new molecules were chemical in nature. Yet they took on some of the behaviors of simple life-forms.

The First Cell

Biochemists say that somewhere along the line, following millions of years of random mixing of organic molecules in the primordial soup, the first cell appeared. (The cell is the smallest unit of life, a collection of molecules that have bonded together.) No one knows how this first cell formed. Two things, however, are likely: It lived in an ocean with no predators, and it was surrounded by nutritious molecules. Robert Hazen says that this cell's offspring "probably filled the world's oceans, consuming much of the organic raw materials and greatly reducing the chance that any other type of cell would spontaneously arise."

Biochemist Richard Dickerson agrees. He thinks that the first living organisms "were presumably one-celled entities resembling modern fermenting bacteria. They would have been scavengers of the organic matter produced by electric discharges and ultraviolet radiation."

Development of Photosynthesis

As these organisms consumed the remaining primordial soup, they needed a new food source. The development of photosynthesis, the process in which a cell converts sunlight into energy, answered that need. Harvard astrophysicist Eric Chaisson says, "Photosynthesis freed the early life forms from total dependence on the diminishing supply of organic molecules in the oceanic broth. In time, the [life-forms] evolved to become all the varied plants now strewn across the face of our planet."

Few fossils remain to tell scientists of the single-celled organisms from the organic soup era. This is probably because geological activity, such as earthquakes, volcanic eruptions, and the slow movement of continents destroyed most of the earth's rocks from that time period. Also, the single-celled organisms were too soft and small to leave much of an imprint in the rocks. Nevertheless, scientists do have a few fossils from that era showing the existence of single-celled organisms.

The development of photosynthesis and the appearance of multicelled organisms marked a time of

Geological activity, including volcanic eruptions, has destroyed most of the rocks that may have contained fossils of early single-celled organisms.

radical change for the planet. Biochemist Richard Dickerson notes, "The next two billion years [following the development of photosynthesis] was a revolution in the nature of the atmosphere of the planet." It went, he says, "from an atmosphere with little or no free oxygen to one in which one out of every five molecules is oxygen."

Life-forms breathed in carbon dioxide and breathed out oxygen. This process led to the formation of an ozone layer in the upper atmosphere. This ozone layer sharply reduced the ultraviolet radiation at the earth's surface. The process of photosynthesis had now replaced the processes of chemical change. At this point the era of chemical evolution faded, and the era of biological evolution began. Dickerson notes, "The pattern of life driven by solar energy was fixed for all time on our planet, and the stage was set for true biological evolution."

To supporters of chemical evolution, the theory neatly explains many mysteries about the beginning of life. Most scientists have adopted it as the best theory. It is not, however, the only explanation.

Primordial Soup Idea Contested

A few scientists object to the idea of the chemical evolution of life. Astronomer Fred Hoyle, for one, does not believe the primordial soup idea is feasible. He says that no one has absolutely proven that life could come from such a mixture.

Hoyle questions how certain important molecules could match up properly in the vast expanse of the oceans. The odds against successful bonding, he says, are extremely high. Without this bonding, chemical evolution would never begin.

Most supporters of chemical evolution, however, have stayed with the theory because of a famous experiment conducted in 1953 by chemists Stanley Miller and Harold Urey at the University of Chicago.

Miller and Urey wanted to see what natural

"The first billion pages in the book of the earth's history are almost completely missing. One must try to reconstruct them from information in the later pages and from what one knows about the other planets and about chemistry in general."

Richard Dickerson, biochemist

"Why do biologists indulge in unsubstantiated fantasies in order to deny what is patently obvious, that the 200,000 amino acid chains, and hence life, did not appear by chance?"

Fred Hoyle, British astronomer

In 1953 chemists Harold Urey (pictured) and Stanley Miller conducted a laboratory experiment that provided evidence for the theory of chemical evolution.

processes might have formed the complex molecules of the primordial seas. Essentially, they tried to duplicate the earth's early conditions—in a jar. First they poured water into the jar, then added ammonia, methane, and hydrogen gases to represent the atmosphere. While they continually heated the mixture, they added electrical sparks to duplicate the effects of lightning. Incredibly, within a few days, the water turned brown. Study of the mixture revealed the formation of amino acids, one of the chemical building blocks of life.

When the experiment was repeated, omitting the electrical sparks, no amino acids formed. This indicated that the acids could only form through a combination of all the ingredients *and* an energy source.

Later, other scientists conducted similar experiments. They used different combinations of gases and used ultraviolet radiation instead of lightning. These experiments revealed results similar to those obtained by Miller and Urey. Each time, the process formed amino acids, various sugars, and other molecules essential to life.

While none of these experiments really created life in a test tube, they showed that life-forming molecules will form when the conditions are right. Today, similar ongoing experiments are successfully duplicating larger and more complex molecules. These experiments seem to add weight to the theory of chemical evolution.

Hoyle, however, remains skeptical. He says, "It is doubtful that anything like the conditions which were simulated in the laboratory existed at all on a primitive earth." Nor did they occur, he says, "for long enough times and over sufficiently extended regions of the earth's surface." In other words, he warns that what happened in a small jar may not necessarily have happened in the much larger laboratory of the earth's ancient oceans.

Biological Evolution

Most scientists accept biological evolution as fact. That is, they believe that the forms of life that exist now developed from the first cell and the mul-

According to the widely accepted theory of biological evolution, these primitive marine life-forms are our ancestors.

ticelled organisms that followed. Scientists can trace this biological change through the fossil record. That record, they say, shows a clear pattern of change from simpler forms to more complex, over billions of years. Thus, all life on earth is related, from the smallest microbe, to the dinosaurs of the past, and the animals, plants, and people of the present.

A few scientists, however, do not agree on how this process first started. Some of them have an explanation different from that of chemical evolution. It is panspermia, the seeding of the earth from outer space.

Panspermia: Life from Space

Panspermia (Greek for "seeds everywhere") was first proposed as a scientific theory in 1903 by the Swedish Nobel Prize-winning chemist Svante Arrhenius. He suggested that life originated not on earth, but on distant worlds in outer space.

P.H.A. Sneath, of the University of Leicester, England, explains the theory: "Arrhenius suggested that life arrived on earth in the form of germs, such as the spores of microorganisms." Such microscopic spores, or "seeds," would have traveled through space from other parts of the universe. Sneath adds that the spores were probably propelled through space "by the weak but continued pressure that is exerted by light rays" on small particles.

When some of these "cosmic seeds" reached the ancient earth, they settled into the atmosphere and eventually fell to the oceans. There, finding the conditions favorable, they began to grow and reproduce. They started the process of evolution. According to this theory, all present life on earth descended from these extraterrestrial (otherworldly) spores.

Fred Hoyle and his colleague Chandra Wickramasinghe are supporters of panspermia. They say that the earth, with its vast oceans, is the perfect "assembly station" for wandering extraterrestrial microbes. Ideal temperatures, a thick, protective

"In essense, the first cell, once it appeared, preempted other possibilities of life."

Robert Hazen, professor of earth science, George Mason University

"This [chemical evolution theory] simply does not fit the facts."

Fred Hoyle, British astronomer

An artist depicts earth's collision with a giant meteor. Perhaps the seeds of life came to earth from the core of a meteor.

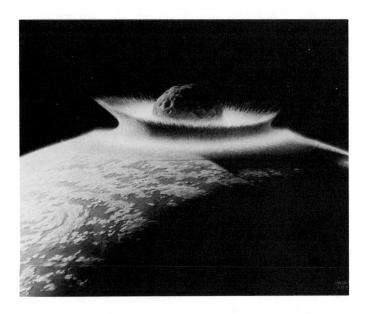

"[Solar] radiation pressure, which flings the tiniest particles into space with enormous velocity, must have played an important part in the transmission of living organisms from one planet to another."

Svante Arrhenius, Swedish chemist, 1903

"One drawback to Arrhenius's panspermia theory is the problem of explaining how organisms could have survived the lethal effects of radiation on their long journey through space."

John Reader, author, *The Rise of Life*

atmosphere, and abundant water and mineral supplies make it an inviting seeding ground for microorganisms from outer space.

Panspermia Criticized

Most scientists have serious reservations about the panspermia theory. Space scientist Joshua Lederberg, for instance, questions how spores could overcome the gravity of their own planets to get into space. He also questions how these spores could survive the deadly ultraviolet rays from the sun or other stars encountered in space.

P.H.A. Sneath is also a critic. He points out that space is so vast that even incredible numbers of spores would become widely scattered. "There would be little chance," he says, "that a planet such as the earth would ever capture any of them. In any case, they great majority would be burned up by the stars or in hot gas clouds."

Hoyle has answered some of these critics by changing Arrhenius's original proposals. Perhaps the seeds of life did not arrive here by floating freely in space. Instead, the spores have been and still are riding through space in the cores of comets or meteors. (Organic molecules have, in fact, been found in the cores of meteorites.) Perhaps these organic molecules were picked up from what Hoyle calls lifeclouds, gigantic clouds of organic material existing in the space between stars. (Such interstellar clouds of organic molecules have indeed been detected by telescope.) Or maybe the seeds originally came from larger bodies in space, from which the meteors or comets broke away eons ago.

Hoyle goes on to suggest that the microbes for certain diseases or even the genes that control our hereditary characteristics came, and continue to come, to earth from such unearthly sources.

Astronomer William K. Hartmann comments, "[This] theory has not been widely accepted but testifies to the continuing interest in the state of

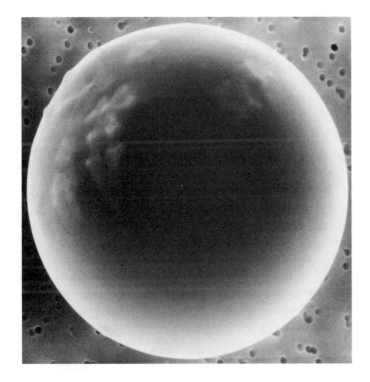

This tiny particle of dust, magnified 15,000 times, is believed to have been shed by a passing comet. Some scientists believe that the spores that produced life came to earth in the same way.

biochemical evolution of material in interplanetary debris from the early solar system."

Hartmann acknowledges that alien microbes might have come by meteorite to earth in the distant past. But, he also believes that these microbes would have been destroyed by the "intense cratering of that era." In fact, he says, it is likely that "rare big impactors"—devastatingly huge meteor impacts— "probably temporarily vaporized the oceans, thus wiping the slate clean of any products of alien evolution." Destruction of alien microbes, he adds, allowed "terrestrial [earthly] evolution to start producing home-grown, true-blue terrestrial life-forms."

Biologist Francis Crick and chemist Leslie Orgel have proposed yet another version of the "life from space" theory. They call their idea directed panspermia. They believe the seeding of space is deliberately planned and carried out (directed) by beings from other worlds.

The biologists say it is possible that advanced civilizations existed elsewhere in the galaxy when

A purported photograph of a UFO. The theory of directed panspermia says that an advanced civilization may have sent a robotic spaceship to plant a primitive form of life on earth.

the earth was formed. One of these civilizations may have deliberately planted a primitive form of life on earth.

Such advanced societies would launch robotic spaceships towards stars likely to have young, primitive planets. Aboard these spaceships would be the spores or microorganisms needed to seed life on other worlds. The ships would protect the spores from deadly stellar radiation during the trip.

Crick and Orgel say, "The spaceship must clearly be able to home in on a star. The packets of microorganisms must be made and dispersed in such a way that they can survive the entry at high velocity into the atmosphere of a planet, and yet be able to dissolve in the oceans."

Space scientist Joseph A. Angelo Jr. discusses another possibility—"that life on earth might have evolved as a result of microorganisms [accidently] left here by ancient astronauts themselves." Of this extraterrestrial garbage theory he says, "It is most amusing to speculate that we may be here today because ancient space travelers were 'litterbugs,' scattering their garbage on a then-lifeless planet."

More Evidence Needed

Skeptics say little evidence supports the idea of directed panspermia. Joshua Lederberg says the theory is weak because there is, so far, no evidence of life on other worlds. Likewise, there is no evidence that alien beings visited the earth in the past. Until such evidence is available, he says, the panspermia explanations will not gain much support.

Most scientists agree that chemical evolution and panspermia remain as the two best explanations for the origin of life. Author John Reader summarizes: "Either life developed from the material[s] . . . of the earth . . . or [it] originated in some other parts of the universe and the earth was, as it were, sown with the seeds of life from space."

Scientists are still looking for the final answer.

"Present-day organisms should be carefully scrutinized to see if they still bear any vestigial traces of extraterrestrial origin."

Biologist Francis Crick and chemist Leslie Orgel

"One can neither prove nor disprove theories of panspermia."

Richard Dickerson, biochemist

Three

Does Life Exist Beyond the Solar System?

From ancient times to the present, humans have looked to the stars and wondered, "Is anyone else out there?" The ancients only knew the stars as pinpoints of light. Yet some people wondered. In 400 B.C., the Greek philosopher Metrodorus of Chios wrote, "It is unnatural in a large field to have only one shaft of wheat and in the infinite universe only one living world." Much later, between 1609 and 1610, the Italian astronomer Galileo Galilei made an important finding. Using a crude telescope, he discovered that the planets were not burning balls of gas, like the stars; they were other worlds. From that time forward, people have speculated that life exists beyond the earth.

With so many thousands of stars peppering the nighttime skies, it is easy to imagine that the universe is home to other beings. Surely other planets must be circling those stars. Surely some of these planets must have life on them. Surely intelligent beings live on some of those worlds. And surely some of those beings are looking at their nighttime skies and wondering the same thing.

Today considerable debate surrounds the question of extraterrestrial life. Many scientists think the universe is the home for hundreds, thousands, or

(Opposite page) The Andromeda galaxy is approximately 2 million light-years away from earth. Is it home to intelligent beings who, like us, gaze out at the nighttime skies wondering, "Is anyone else out there?"

millions of technological civilizations. Others think that few if any other civilizations exist in the universe. Some believe that conditions like those on the ancient earth exist or have existed on countless other worlds. And, they say, where conditions are similar, life will blossom. Others believe that the evolutionary steps toward life and intelligence are filled with many roadblocks. Therefore, the chances of intelligent life arising on another world are exceedingly small.

Principle of Mediocrity

Supporters of widespread life in the universe base their optimism on a scientific concept called the principle of mediocrity. According to this principle, the laws of nature relating to the formation of stars, planets, and life are the same everywhere. If stars and planets are able to form in other parts of the universe, then life should be able to develop there as well. In this view, the earth is a mediocre world. That is, it is typical of planets anywhere in the universe. If life exists on earth, then it must exist elsewhere, on other earthlike worlds.

Astronomer Carl Sagan of Cornell University says, "From our knowledge of the processes by which life arose here on the earth, we know that similar processes must be fairly common throughout the universe." He says it is likely that alien life-forms would develop on other worlds. Occasionally, these life-forms would develop intelligence. And sometimes these creatures would create civilization and a high technology.

Aerospace physicist John Billingham says, "The vast majority of scientists today believe it is likely that extraterrestrial life exists. If there is another star like the sun and going around it is a planet like the earth, which is likely the case, then the odds of life beginning and evolving are high." Billingham formerly headed the SETI (Search for Extraterrestrial Intelligence) project for the National Aeronautics and

"Many civilizations have been generated in our galaxy."

Nicolai S. Kardashev, Russian astronomer

"There is no hard data which permits us to conclude ... that the circumstances for the development of life beyond our system exist."

David Leigh Rodgers, medical director

Space Administration (NASA). He also serves as chief of the Life Sciences Division of the NASA Ames Research Center in Pasadena, California.

Astronomer Carl Sagan believes it highly likely that intelligent life exists elsewhere in the universe.

Evolution of Life Is Rare

Not all scientists take such an optimistic view of life developing on other worlds. Dr. David Leigh Rodgers of the Stanford Medical School says, "The chances for life to form once are infinitely small. From what we know of our solar system and the universe, the chances for it to form again [somewhere else] are zero."

Tulane University physicist Frank Tipler agrees. He says that most evolutionary scientists believe the odds are against an intelligent species evolving from simple one-celled organisms. Because of this, Tipler concludes, "We are probably the only intelligent species ever to exist in the galaxy, and quite possibly the only such species that has ever existed in the known universe."

Evolutionist Stephen Jay Gould, however, disagrees with Tipler's conclusions. He says that Tipler misstates the beliefs of many evolutionists. Gould points out that "at least four quite respectable evolutionists [signed] the international pro-SETI petition released by Carl Sagan." This 1982 petition urged the U.S. government to launch a search for extraterrestrial civilizations. "Evolutionary biologists . . . maintain a diversity of views on this subject," says Gould.

Still, Tipler asserts that extraterrestrials do not exist. He says:

> I base my claim that we are alone in the galaxy on the idea that interstellar travel would be simple and cheap for a civilization only slightly in advance of our own. Thus if a civilization approximately at our level had ever existed in the galaxy, their spaceships would already be here. Since they are not here, they do not exist.

University of Virginia physicist James S. Trefil agrees, saying that

> the extraterrestrials have never been here, and they're not here now. From this we can conclude that there have probably been no advanced civilizations in the history of the galaxy up to this point. This is certainly consistent with the data we have.

Von Neumann Machines

If they did not visit the earth themselves, say Tipler, Trefil, and others, advanced aliens would have built robotic probes to explore the cosmos. The ability to construct such machines would be the result of millions of years of technological development. Called von Neumann machines (after the mathematician John von Neumann, who first proposed them), these intelligent spacecraft would be capable of manufacturing copies of themselves. The

Evolutionist Stephen Jay Gould supports the view that evolution is likely to have occurred in other worlds.

copies would in turn explore additional portions of the galaxy, making additional copies of themselves along the way. In this spread and multiply manner, the von Neumann machines would explore the entire galaxy within 300 million years. Everywhere they went, these machines would leave evidence of their arrival, including mining sites, manufacturing debris, and artifacts of some sort.

No Evidence of Alien Visit

Tipler believes that if advanced intelligent beings existed, they would have come to earth a billion years ago. Since the only life on earth then was one-celled organisms, the aliens would have left evidence of their visit somewhere in the solar system. They would have had no reason to hide their technology. For instance, artifacts left over from mining operations would be found on asteroids (small rocky bodies circling the sun beyond the orbit of Mars). Because no such evidence of alien visits exist, Tipler concludes that extraterrestrials do not exist.

Jill Tarter, former project scientist for NASA SETI, however, sees fault in this viewpoint. "The logic [of the von Neumann machine argument] has a compelling force, but we have several reasons to doubt . . . the inevitability of space travel. Most obvious are the limitations of technology, . . . lack of motivation or desire, and cost/energy effectiveness." In other words, advanced aliens may not want to take the time and trouble to travel long interstellar distances. This could be especially true if the financial costs and energy requirements are high.

Likewise, astronomer and SETI Institute president Frank Drake believes that sending interplanetary probes would be expensive and inefficient, even for an advanced civilization. To succeed, the launching civilization would have to send out millions of probes. The probes would have to work perfectly for thousands, even millions of years. And the home planet would have to wait thousands or mil-

"For the first time in our history there is a reasonable scientific basis for supposing that there is intelligent life out in the galaxy and in the universe."

John Billingham, chief of Life Sciences Division, NASA Ames Research Center

"If we do indeed follow the path of expansion into the universe, then the first (and only) extraterrestrials we will meet will be our own grandchildren."

James S. Trefil, physicist, George Mason University, Fairfax, Virginia

Jill Tarter, former project scientist for NASA SETI, finds fault with the assertion that if advanced intelligent beings existed they would have come to earth a billion years ago.

"I'm sure there's something out there to find."

Jill Tarter, former project scientist, NASA SETI

"I'm skeptical as to whether they're going to find anything."

Robert Rood, astronomer, University of Virginia

lions of years to receive updates from the probes. Such schemes, Drake says, are simply not practical.

According to Drake, there are plenty of other reasons to explain why advanced aliens are not here even if they do exist. Some of these reasons include:

- They see no personal gain in creating a costly army of von Neumann machines.

- They are content to colonize their own star system and leave the [rest of the] galaxy alone.

- They, like us, have found radio communication the more promising alternative, and are in fact engaged in it even as we debate the issue.

So, Drake asks, "Where are they? They are probably living quite comfortably, with a high quality of life, near the planets where their lives began. They are there in great numbers for us to find—with radio transmissions."

Some scientists thought the debate about extraterrestrial life might be settled soon. On October 12, 1992, NASA began searching the skies for signals from extraterrestrial civilizations. However, on October 1, 1993, Congress cut funding for the NASA SETI project.

SETI scientists have long believed that successful detection of extraterrestrial beings may depend on how many advanced civilizations actually exist in our galaxy. Surprisingly enough, one scientist has devised a formula for estimating the number.

The Drake Equation

In 1965 Frank Drake devised this equation:

$$N = R* \cdot f_p \cdot n_p \cdot f_l \cdot f_i \cdot f_c \cdot L$$

The formula is not as complicated as it looks. On the left side, the "N" represents the number of technological civilizations in the galaxy. On the right side, scientists insert figures for: the average yearly rate of star formation in the galaxy ($R*$); the fraction of stars having planets (f_p); the number of suitable planets per planetary system (n_p); the fraction of planets on which life starts (f_l); the fraction of life that evolves to intelligence (f_i); the fraction of intelligent species that develops the ability to communicate across space (f_c); and the length of time, in years, that a civilization remains detectable (L).

The answer to the Drake equation depends on the figures inserted into it, and none of the elements of the equation are known facts. Therefore, scientists optimistic about the prospects of extraterrestrial life will usually insert higher numbers. Scientists pessimistic about the prospects will insert lower numbers. Either way, the formula involves a lot of guesswork.

Carl Sagan joins Drake in believing the equation shows that the universe is teeming with intelligent life. He says, "Our best guess is that there are a million civilizations in our galaxy at or beyond the earth's present level of technological development." He adds that if these civilizations are randomly distributed throughout space, the distance between the earth and the nearest civilization would be about three hundred light-years. That is about 1,800 trillion miles away.

Drake himself says, "The basic premise behind the equation is that what happened here [the evolution of intelligent life] will happen with a large fraction of the stars as they are created, one after another, in the Milky Way galaxy and other galaxies." People who think the equation is simply a guess are wrong, he says. "It is just the opposite, since the [events] it assumes to take take place in the universe are only those we are sure have taken place at least once."

Drake is modest about the equation. He says, "It expressed a big idea in a form that a scientist, even a beginner, could [understand]. . . . It amazes me to this day to see it displayed prominently in most textbooks on astronomy."

Space consultant Frank White, author of *The SETI Factor*, takes a middle approach to the equation. He says, "It is not an equation with a clear answer, because it contains far too many unknowns. It should be seen perhaps as a model, a simplified way of thinking about the most important SETI questions."

Reliability of Equation Questioned

Others question the reliability of the equation altogether. Stanford University astrophysicist Ronald Bracewell is one. He says, "Drake's equation is obviously wrong. . . . It's trying to estimate how many civilizations there are, something that we don't know and that we will have to arrive at by observation." The fact that the equation depends on so much guesswork, he says, makes it meaningless.

Frank White, however, believes that time will prove the equation's value. He says, "Over the centuries humans will transform the Drake equation from a model filled with assumptions to a description of reality in our galaxy and the universe."

The NASA SETI project, some scientists thought, might have proved the value of Drake's

"I find nothing more tantalizing than the thought that radio messages from alien civilizations in space are passing through our offices and homes right now, like a whisper we can't quite hear. . . . Information-laden radio messages are the quarry we seek in the search for extraterrestrial life."

Frank Drake, astronomer, SETI Institute

"A radio signal beamed toward an uninhabited stellar system is wasted effort, whereas a probe would at least send back information about the system."

Frank Tipler, physicist, Tulane University

Astronomer Frank Drake developed a mathematical formula to determine how many civilizations might be in outer space—and the chances of finding them.

equation. Detection of a signal from another civilization would have ended all the arguments and could have opened the doors to a new vision of humanity's place in the universe.

The NASA SETI Project

NASA planned to use two techniques to search the heavens for signs of extraterrestrial intelligence. The first was a targeted search examining about eight hundred to one thousand sunlike stars, all within about eighty light-years of earth. In this search scientists were to look for weak or stray radio signals beamed from individual planets circling those stars. The search was to take place over a ten-year span.

The second technique involved a sky survey; that is, scanning the sky in checkerboard patterns, looking for powerful radio-wave beacons. The signals NASA was to look for would have been beamed by supercivilizations that had migrated from their home planets and used the beacons as navigational aids. Scientists believe there are several reasons these migrations might take place. A civilization might need additional living space or more raw materials. Or it might want to conduct scientific explorations of other worlds. The sky survey was to take place over a five- to seven-year period.

Both searches involved radio telescopes, computers, and other sophisticated equipment to detect extraterrestrial signals. The radio telescope is similar to the satellite dish antenna used to receive television signals. It detects radio signals over wide frequency ranges. Working with the radio telescopes are computer-linked multichannel spectrum analyzers (MCSAs). They scan millions of frequencies at the same time. If the computer system detects an unusual signal, it alerts scientists to take a closer look.

All chemical elements send out radio waves at different frequencies. This adds up to millions and millions of frequencies. In designing the SETI proj-

Radio telescopes, which can detect radio signals over wide frequencies, are used in the search for extraterrestrial intelligence.

ect, NASA's biggest problem was deciding which frequencies to scan to find purposeful signals. NASA's scientists decided to center their signal search around the narrow microwave frequencies of the element hydrogen (H) and the compound hydroxyl (OH). The combination represents water (H_2O). Because of this the frequency range is known as the water hole.

Water Hole Frequencies

Scientists had several reasons for selecting water hole frequencies. First, this part of the electromagnetic spectrum is fairly clear of interference or "noise" from natural objects in space. For that reason it is an ideal bandwidth for the sending and receiving of radio signals. Bernard Oliver was deputy chief of NASA SETI. He says that radio observers in other parts of the galaxy would recognize the same noise-free part of the spectrum. They too "would conclude that this was the best part of the spectrum for interstellar communication." Second, water is a vital ingredient of living things. Selection of water hole frequencies symbolizes the recognition of that fact. Advanced aliens would probably recognize the same symbolism. They would probably turn to the water hole frequencies for that reason as well.

Why do scientists think they might be able to distinguish a signal that is of intelligent origin? Sam Gulkis of the NASA Jet Propulsion Laboratory explains, "The only thing that makes sense is to assume the signals we're looking for are intentionally built [by aliens] so [that] they are easy to detect."

Gulkis compares aliens trying to signal us to humans trying to get a dog's attention:

Clapping will work because the sound is different enough that it won't blend in with the background. A whistle is loud enough to rise above the background. An alien signal ought to be something similar—a pulse or a strong tone—so we can distinguish it from the noise.

"The incredible improbability of alien intelligence should be taken into account when deciding how much of our effort SETI should occupy."

Zen Faulkes, University of Victoria, British Columbia, Canada

"I believe the promise of SETI is far greater than its perils. It represents the highest possible form of exploration, and when we cease to explore, we will cease to be human."

Arthur C. Clarke, science fiction author

In other words, these scientists think extraterrestrials are trying to get our attention. Like a dog recognizing its owner's voice, or a friend recognizing another's shout in the crowd, we would recognize the signal of other intelligent beings.

What Are the Chances of Success?

The successful detection of an alien signal could take years, or it could happen quickly. Our own galaxy alone has about 200 to 400 billion stars. Scientists figure we would have to scan about 200,000 of them just to have a chance of detecting another civilization. Before the funding was cut, physicist and SETI equipment designer Kent Cullers described the NASA project as "searching the equivalent of the *Encyclopaedia Britannica* every second to find the part that says 'Hi, we're the aliens.'" He added, "We will not by any means have surveyed the entire galaxy when we finish the ten-year search. This is the first step along what may be a very long road."

Some scientists and politicians have questioned the wisdom of spending time and money on SETI and these questions ultimately led to the project's cancellation. Democratic senator Richard Bryan of Nevada says the money being spent in this way is needed for more urgent problems. These problems include large budget deficits, health care needs, and poor levels of educational funding. Bryan says, "the federal government has no business financing something as superfluous [unnecessary] as [SETI]."

Republican congressman Silvio Conte of Massachusetts is even less kind to SETI efforts. In 1990 he told fellow legislators, "Of course there are advanced civilizations in outer space. But we don't need to spend six million this year to find evidence of these rascally creatures. We only need seventy-five cents to buy a tabloid at the local supermarket."

Furthermore, Howard University philosopher and SETI critic Edward Regis questions whether

"The probability of success is difficult to estimate, but if we never search, the chance of success is zero."

Giuseppe Cocconi and Philip Morrison, physicists, Cornell University

"If SETI does proceed as planned, I propose we establish a program called SCOTI, the Search for Congressional Intelligence."

Unnamed congressman, 1990

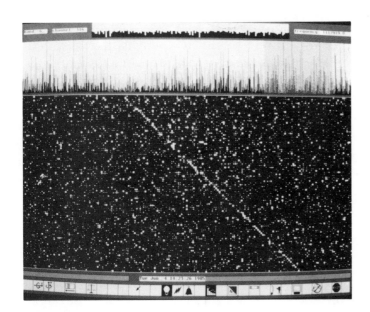

A computer screen displays a signal from the *Pioneer 10* NASA spacecraft, detected by SETI equipment. SETI scientists hoped that their equipment would one day detect a signal from extraterrestrial beings.

finding a signal will be useful. "If the senders are as dissimilar from us as we are from ants," he says, "we have no reason to suppose that their transmissions will encode anything we can recognize, much less use."

SETI scientists have disputed these criticisms. They say that we have plenty to gain for the time and money spent.

How Can We Benefit?

Supporters of the NASA project believe that the discovery of extraterrestrial signals would pay off in many ways. Science writer John M. Williams says, "Beyond scientific merit, many now recognize three social values to SETI. One, it is a way to perhaps look at how civilizations solve their technological crises." In other words, civilizations that have avoided war and environmental destruction may have something to teach humanity. "Two," he says, "it unifies many scientific disciplines into a single story of the universe." Scientists from many different fields may finally agree on a single version of

"The detection of a radio signal from an extraterrestrial civilization would be a transforming event in human history."

David W. Swift, sociologist, University of Hawaii

"We will learn more about the existence of extraterrestrial intelligence if we spend money on studies of the evolution of life on earth, rather than on fanciful radio searches."

Frank Tipler, physicist, Tulane University

how life formed in the universe. "And three, it will generate direct scientific spinoffs, like new computer chips designed to perform complex mathematical functions twice as fast as any chip previously on the market."

New Information Could Benefit Humanity

Carl Sagan notes that the extraterrestrials would probably view the universe in a different way than we do. They would have different art, music, politics, philosophy, science, and social values. They would know things that humans have never thought of. Discovery of this new information, Sagan says, would teach humanity important lessons. For instance, it would show that the universe does not belong to humanity alone. The human view of the universe would no longer be the only view. Sagan believes exposure to alien ideas and knowledge would help humanity grow in many ways.

Jill Tarter agrees with Sagan. She believes that the cost of SETI would be "well worth the scientific and cultural riches that could be received from an advanced civilization."

Frank Drake also agrees that we could benefit from SETI. He says:

> I fully expect an alien civilization to bequeath us vast libraries of useful information. . . . This *Encyclopedia Galactica* will create the potential for improvements in our lives that we cannot predict. . . . Another Renaissance will be fueled by the wealth of alien scientific, technical, and sociological information that awaits us.

But physicist Frank Tipler says those who believe that an extraterrestrial message will help us in some way are misleading themselves. This belief, he says, is similar to the belief held by some people that UFOs have come to earth to save humanity from a variety of problems, disasters, and blunders.

Most scientists believe that receiving an extraterrestrial signal would have an immediate impact on our lives. NASA's John Billingham says this impact would be equivalent to the changes brought about when, near the end of medieval times, people learned that the earth was not the center of the universe.

Public Interest Could Fade

NASA's Bernard Oliver believes that people would at first be excited by the news that life exists on another world. But after a while the excitement would die down. This would happen because it would take a long time to get additional information from the senders. For example, if scientists received a message from a planet circling the nearest star, it would take four years for the aliens to receive our reply. That is because radio messages travel at the speed of light, and the nearest star is four light-years away. Then, it would take another four years for the aliens to send their response. Eight years would have passed since detection of the original signal. For stars farther away, the reply-and-response times would be even greater. Thus, Oliver believes, public

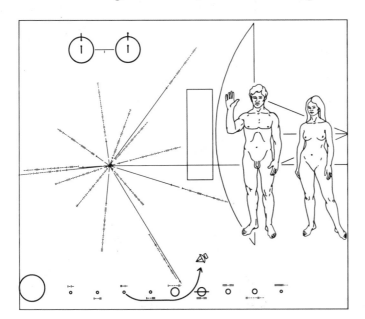

The *Pioneer F* spacecraft, sent outside of the solar system, carried this plaque. Scientists hoped it would answer the question "What do earth beings look like?" for any alien civilization that might intercept the spacecraft.

interest in the signal would soon fade. "People would go back to the National Football League," he says, "and life would go on as normal." But as scientists continued to study the original message, there could be a trickle-down benefit of useful information for all humanity.

Edward Regis, however, believes that even an information-laden message would have limited effects. At first, people might be amazed by new wonders of the universe as revealed by an advanced civilization. But after a while people would return to what is most comfortable for them—the routines of *human* daily life. Regis concludes, "Extraterrestrials might always be no good for us."

What If We Do Not Detect a Signal?

Supporters saw value in the SETI project even if scientists failed to detect civilizations in space. Carl Sagan comments, "It would speak eloquently of how rare are the living beings of our planet and would underscore, as nothing else in human history has, the individual worth of every human being." In other words, the realization that the earth is the only life-bearing world in the universe might encourage people to treat one another better. People might learn to take better care of the earth and work harder to preserve the life-forms it hosts.

But Edward Regis doubts there is much to gain from lack of contact. He says:

> Failure to find something out there would not entitle us to conclude very much. If we make a search and end up only with absence of evidence of extraterrestrial intelligence, then the only thing we would be entitled to conclude is that we have no evidence of their absence. In other words, we are no better off—and no worse either—than we were when we started. We still don't know if they're out there.

Although funding for SETI has, for now, come to an end, debate over its merits will probably arise

NASA's Bernard Oliver believes that the detection of an extraterrestrial signal would have little effect on our daily life.

again. Many scientists still believe it is a worthwhile project. Some members of Congress may also share this view. But the nation has many pressing problems and Congress is under great pressure to spend money more wisely than it has in the past.

Perhaps someday NASA's search for extraterrestrial life will resume. Thomas R. McDonough believes it should go on. "Nobody on earth knows whether we're alone or if the universe is buzzing with life," he says. "Sticking our heads in the ground never got us anywhere in science. If we don't search, we'll never find anything. If we do search, the least we'll do is explore the universe."

The answers wait—in the vast cosmos.

Supporters of SETI believe that the failure to detect an extraterrestrial signal would underscore how precious earth and human life are.

Four

Is Mars the Site of Ancient Cities?

For centuries the planet Mars has fired the human imagination. To many cultures, the planet's red color suggested images of blood and fire. Thus the ancient Sumerians, Greeks, Persians, and Egyptians named the planet for their gods of war. The early Romans did, too, and it is the Roman name that we use today.

The Polish astronomer Nicholas Copernicus, the German astronomer Johannes Kepler, and the Danish astronomer Tycho Brache were among the first to realize that Mars was not merely a flickering star but another world. The Italian astronomer Galileo Galilei confirmed this when in 1610 he became the first to view Mars through a telescope. Unfortunately, the view was extremely fuzzy. All Galileo could learn about Mars was that "it was not perfectly round."

In 1659 the Dutch astronomer Christiaan Huygens used a telescope with much better lenses. He became the first to see dark markings on Mars. He also determined that Mars rotated on its axis like the earth. He figured that the Martian day was close "to twenty-four terrestrial hours." It is, in fact, twenty-four hours, thirty-seven minutes, and twenty-two seconds long. In 1672 the Italian astronomer

(Opposite page) Mars has long excited the human imagination. Some people believe it was once home to an ancient civilization.

German astronomer William Herschel, who was the first to correctly identify the white patches on Mars as ice caps, believed it likely that the planet was inhabited.

Giovanni Cassini became the first to see white polar caps on Mars.

In the 1870s the German astronomer William Herschel, living in England, began observing Mars. He, too, saw the white patches covering the polar regions. In 1874 he correctly identified the patches as ice caps. He also noticed that the caps changed in size and shape with the seasons. He discovered that Mars "has a considerable but moderate atmosphere." The dark areas seemed to be great oceans. It was entirely possible, he thought, that Mars was inhabited. He concluded that the Martians "probably enjoy a situation in many respects similar to our own."

A Civilization on Mars?

A few years later an Italian astronomer, Giovanni Schiaparelli, reported seeing strange markings on the surface of Mars. He called these markings *canali*, Italian for "channels." When the American astronomer Percival Lowell read of this observation, he thought that Schiaparelli had seen *canals*. Lowell assumed these were artificial waterways constructed by civilized Martians. Lowell believed that if he could confirm the existence of these canals, he could prove that another civilization existed on Mars.

Telescopes at the time still revealed Mars as a fuzzy image. Thus, in 1877, when Lowell pointed his telescope toward Mars, he struggled to make out details on the planet's surface. He saw the large dark areas observed by earlier astronomers. Lowell thought these areas were not oceans, but areas of vegetation. Then he saw what he was looking for: faint lines crossing the areas between the dark patches. Surely these were great canals constructed by intelligent beings. These canals would carry water across the deserts from the melting ice caps to

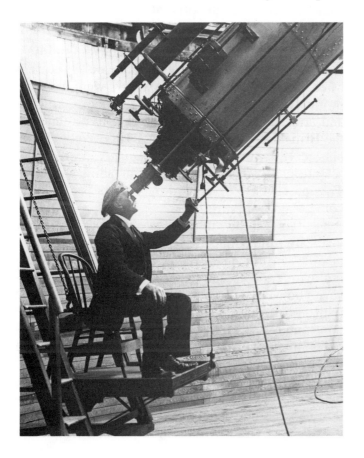

Astronomer Percival Lowell detected lines on the surface of Mars through his telescope. He became convinced that the lines were canals constructed by intelligent beings.

the vegetation. Lowell drew numerous maps of the canals and wrote books about civilizations on Mars.

Many astronomers studied Lowell's maps and turned to their own telescopes to search for the canals. Some reported seeing them and drew their own maps. Others could not see the canals at all. Thus not all astronomers were convinced they existed.

Later, better telescopes failed to detect the canals. Scientists concluded that the lines seen by Lowell were optical illusions. Still some scientists had reason to think life existed on Mars. Many still believed the dark patches represented vegetation. The fact that the patches grew and shrank with the seasons seemed to support that conclusion.

A War of the Worlds

Lowell's theories quickly caught the public's fancy. Imaginations soared. Many believed that Mars was home to life. Many writers were inspired, promoting the idea of Martian life in books, articles, and novels. The most famous writer to do this was the British science fiction author H.G. Wells. In his classic 1898 novel *War of the Worlds*, he wrote of scientifically advanced Martians invading the earth. This plot would be copied by writers for years to come.

In fact, on the night of October 30, 1938, forty years after the publication of Wells's novel, listeners heard a startling announcement on CBS radio. Martian invaders had landed at Grovers Mill, New Jersey. The Martians were fighting ferocious battles with the army. As the regularly scheduled program continued, repeated bulletins and on-the-spot reports interrupted the show. Martians were landing everywhere. Thousands of people were being killed by deadly heat rays and poisonous smoke. Cities were on fire. The Martians were destroying everything, just as they had in the H.G. Wells novel.

Many listeners panicked. Some fled their

homes. Some covered their faces with wet handkerchiefs for protection against smoke. Others fled in cars to escape the Martian army. Emergency calls swamped the police and military. Hospitals treated people for shock and hysteria.

The radio broadcast was, of course, just a play, produced by a young actor named Orson Welles (no relation to H.G. Wells). It was a dramatic, updated enactment of the Wells novel. It was announced as a play at the very beginning of the broadcast and several more times during the show. But many people were too excited or frightened to notice. They thought the bulletins were real.

The panic showed that the idea of life on Mars was firmly fixed in the minds of many people. But as scientists collected more information about Mars, the planet seemed less and less capable of supporting life. In 1963 astronomers at Mt. Wilson Observatory, near Los Angeles, took infrared measurements of the Martian atmosphere. The measurements revealed that the atmosphere was composed

Thousands of people thought the *War of the Worlds* broadcast, narrated by actor Orson Welles, was a live report of an actual Martian invasion.

mostly of carbon dioxide with only a trace of water. Still scientists expressed hope that some form of life existed on Mars.

Satellite Discoveries

In the mid-1960s the United States at last had the technology to launch satellites into space. As part of a bold new plan to explore the planets, NASA decided to take a closer look at Mars. On July 14, 1965, the *Mariner 4* satellite passed within six thousand miles of Mars. It took twenty-two photographs and beamed them back to earth. Scientists were expecting to see a world somewhat like earth. They were shocked to see instead a heavily cratered surface. There were no canals, oceans, or patches of vegetation. Hopes for finding a civilization on Mars were shattered.

Some scientists still believed it was possible that simple life-forms, such as moss or lichen, lived on the surface of Mars. Others thought that small plants or animals might have lived there in the past but had died out. With this in mind, scientists designed new spacecraft that could land on Mars and sample the soil for signs of life.

On July 20, 1976, *Viking 1* landed in an area called the Plain of Chryse. On September 3, 1976, *Viking 2* landed in an area named the Plain of Utopia, more to the north. Both landers transmitted ground-level photographs and other information back to earth. Orbiters for both craft continued to circle the planet, taking thousands of pictures.

Ground-level photographs revealed a desert strewn with rusty-red rocks and boulders. A few thin, wispy clouds floated high in a red sky. Measurements confirmed a cold, thin atmosphere, unfit for living things. No forms of life were visible in the photographs.

Each lander conducted three life-searching experiments. Instruments aboard the landers heated soil samples, looking for chemical reactions that

would be produced by microscopic life. Scientists were excited when some of the experiments produced the hoped-for reactions. Other scientists, however, offered a different explanation for the results. They said that unusual chemicals in the Martian soil caused the reactions. Eventually, many scientists agreed that the unusual soil probably did cause the chemical reactions. Some disagreed, saying the experiments did not prove things one way or another.

Detailed photographs from the Viking satellites circling Mars revealed a planet of many natural wonders. In one area a great cluster of volcanoes sits atop a tremendous bulge in the planet's surface. One of the volcanoes, Olympus Mons, is six times the height of Mount Everest. Scientists believe Olympus Mons is the largest volcano in the solar system. On another part of Mars, the planet's surface is gouged by Valles Marineris, a canyon four

An artist's conception of the *Viking 1* satellite in orbit around Mars. Viking photographs shattered hopes for finding a Martian civilization.

A Viking photograph of the cliff surrounding Olympus Mons, which scientists believe is the largest volcano in the solar system.

times deeper in some places than the Grand Canyon and nearly three thousand miles longer. In addition to craters, countless mountains, valleys, bluffs, and mesas cover Mars's surface. Most of these landforms are shaped and formed by high winds and dust storms.

The Viking photographs also revealed that Mars was probably once a very different world. They showed evidence of ancient flood basins, rivers, and tributaries. Some areas even looked like the dried-up beds of ancient seas. Could Mars once have had a thicker atmosphere? Was there once running water, maybe even oceans, on the surface? Could Mars have once supported life after all?

Among the scientists studying the Viking photographs were Vincent DiPietro and Gregory Molenaar of the Goddard Space Flight Center in Maryland. Both were affiliated with NASA projects. In 1979 DiPietro and Molenaar were sifting through thousands of the photographs at the National Space Science Data Center files. Amid all these photographs was frame number 35A72. It was taken over an area called Cydonia. Something unusual immediately caught the attention of the two scientists.

Staring straight up at them, from the top of a fifteen-hundred-foot-high, mile-wide mesa, was the unmistakable shape of a human face.

Discovery of the Face

DiPietro and Molenaar were not the first to see this face in the rock. It was first noticed in 1976 when the picture was beamed back to earth by the Viking orbiter. As the image formed on the screen at the Jet Propulsion Laboratory (JPL) in Pasadena, California, scientists chuckled in amusement. Here, indeed, was a face, featuring eyes and eyebrow ridges, a mouth, and a chin. Across the top of the head and down the sides was what seems to be a headdress of sorts, giving the face an appearance something like that of an ancient Egyptian.

Planetary astronomer David Morrison describes the face further:

> The Mars face is a squarish low hill or mesa about one mile across located in a region known as Cydonia. When seen from above under oblique late-afternoon lighting, the eroded top of this hill bears a striking resemblance to a human face. . . . [It is] part of an extensive area of isolated wind-eroded hills.

Many of these hills, often in the rough shape of pyramids, would be unusual on earth.

To the scientists at the JPL, the face was an amusing trick of light and shadow but of little interest otherwise. Journalists and science writers present at the Jet Propulsion Laboratory accepted this explanation.

Copies of the picture and an enlargement of the face area were later distributed to the press. Morrison says, "The image [of the Mars face] was released to the press as a sort of joke—an example of the tendency to recognize apparent anthropomorphic [human] features in an exceedingly complex and alien landscape."

Over the next few weeks, some reporters began

"My own impressions [of the face] were that the photo confirmed the over-all symmetry of the head. Including its supporting structure, the face was also shown to be framed completely around, as if wearing a helmet with a heavy, armored chin strap."

John Brandenberg, physicist

"I don't think there is anything to it. There are many photos taken of Earth, from satellites, which look like that, too, and we know what they are: they are natural formations."

Melvin Calvin, chemist, University of California, Berkeley

to think the face might be the work of intelligent beings. In response Viking project scientist Gerry Soffen told them that a second photograph of the face area was taken several hours after the first. This time, however, lighting conditions were different. In this picture, he said, the face had vanished: "Isn't it peculiar what tricks of lighting and shadow can do? When we took a picture a few hours later it all went away; it was just a trick, just the way the light fell on it." Following this announcement, scientific and public interest in the face dwindled. The face in the rock was soon forgotten.

Scientists Investigate Further

But in 1979 DiPietro and Molenaar's curiosity was aroused. They decided to look further to see if they could find the second picture of the face. Sure enough, after searching through thousands of photographs, they found another picture, frame number 70A13. But it was not taken just a few hours later, as reported by NASA. It was taken *thirty-five orbits* later, at an angle twenty degrees higher than the first picture. (This particular orbit occurred a month later. It represented the next time the satellite passed over Cydonia at a low enough altitude to get a good picture.) There, as before, was the same face in the rock, staring upward toward space.

David Morrison accepts NASA's explanation of the face as an optical illusion. But he believes officials handled communication about the second photo in an embarrassing way. The officials, he says, only wanted to avoid the possibility of sensational publicity: "While plausible, the [NASA] story was untrue. A few hours after the original picture was taken, the face was on the night side of the planet and could not have been seen; in fact, the area was not rephotographed for another month." Morrison calls the incident "regrettable" and says that "such carelessness provides chaff to fuel paranoid fantasies about government coverups."

"[The face shows] established mathematical proportions for human images—ratios of distances between the forehead, eyes, nose, mouth, and chin."

Richard Hoagland, Mars Investigation Team

"A mythology has grown up around the Face, and whole books have been written about it. But most people find that a close study of the photographs—even a casual study, for that matter—shows the face to be just the side of a mountain."

Frank Drake, astronomer

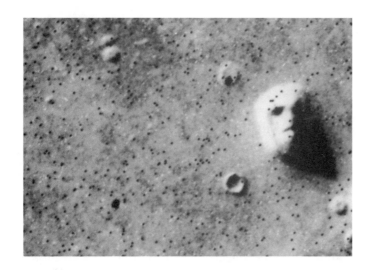

This rock formation (right) photographed by the Viking satellite resembles a human face. While it is probably only an optical illusion, some people speculate that it is a stone monument built by an ancient civilization.

DiPietro and Molenaar used computerized procedures to improve detail in both pictures. After considerable study they concluded that the face was a real structure, not just an optical illusion.

Martian Pyramids

In studying frame 70A13, the two scientists noticed a cluster of pyramids not far from the face. Especially interesting was a gigantic pyramid, about 1 by 1.6 miles in size. They observed that "there appear to be four sides that go down to the surface at sharp angles. The corners exhibit symmetrical material, almost as if they were being buttressed" (artificially supported). Some might suggest that this material was caused by erosion. But, say the scientists, "One would expect erosion at the center of the wall rather than at the corners."

DiPietro and Molenaar observed that the face's features are balanced. That is, the eyes, cheekbones, and helmet are in the correct positions, as in a real face. This fact, say the scientists, combined with "the non-uniformity of alignment between adjacent pyramids . . . leave doubt that nature was totally responsible." In other words, someone built these structures.

In 1980 DiPietro and Molenaar released their findings in a scientific paper entitled "Unusual Martian Surface Features." In it the authors suggested that the face and nearby pyramids were artificial structures. They distributed their paper at various professional meetings. Most scientists were either openly hostile or politely cool to the paper's suggestions. They said that people have a tendency to see faces in many different things, such as clouds and rock formations.

Evidence of a Lost Civilization?

Science writer and *Monuments of Mars* author Richard Hoagland, however, believes the face and other Cydonia objects are strong evidence of an ancient, lost civilization. In fact, he believes that the nearby pyramids are the remains of a city. He points out that there even appear to be grid patterns similar to those of city streets.

Hoagland mentions other objects that also suggest an ancient civilization. One of these is the Fortress. The Fortress features straight, towering walls and an apparent interior section. The Cliff has behind it an area of rugged terrain that suggests diggings. There is also an inclined feature that Hoagland calls an access road.

Hoagland says that one five-sided pyramid is constructed to unusual proportions. The pyramid is shaped on the top somewhat like a five-pointed star. The points, says Hoagland, are similar in proportion to those of a human figure with legs, arms, and head outstretched. To Hoagland, this is evidence of construction by humanlike beings. He named this pyramid the D&M Pyramid, after DiPietro and Molenaar, who discovered it.

All of this, Hoagland says, is evidence that *someone* built the various objects. He offers three possibilities of who those people were. First, these objects may have been constructed by "indigenous Martians." That is, the objects were built by a race

"The Face does not belong at Cydonia because it is humanoid. . . . The Face must have been put there by people who had significant prehistoric contact with us or who are actually related to us."

Richard Grossinger, publisher

"At no time in the six-year investigation and analysis period associated with the *Viking 1* and *2* explorations did NASA ever believe the 'faces' were anything but a curious formation caused by wind processes."

Charles Redmond, NASA public affairs officer

of beings who once inhabited the planet but have since disappeared. Second, the objects may have been built by designers "from beyond the solar system." At one time, these alien beings visited the earth and then visited Mars. They built the structures on Mars, similar to structures seen on earth, to mark their visit. Third, an ancient, advanced race of beings from earth went to Mars and built the objects. Hoagland does not prefer one explanation over another. But he points out the similarity between the Martian structures and ancient human structures on earth. To him, it is evidence of a possible connection between the two planets in the distant past.

Hoagland believes that the Cydonia objects were constructed to mark a rare alignment of the earth and sun, as viewed from the Cydonia area. (Alignments occur when bodies in space, such as planets or stars, seem to line up in a certain way. The alignments occur in appearance only. The illusion of lining up is caused by the changing positions of planets as they circle the sun.) In this case Hoagland describes the hypothetical Martians, standing in a city square half a million years ago. "They would have seen the earth rise brilliant in the

Richard Hoagland (above) believes it possible that alien beings who first visited earth traveled to Mars and built pyramids similar to those built by the ancient Egyptians (left).

"There is a striking relationship between this 'city' and 'the face.' The placement of the city is at a perfect right angle to the face, facilitating a perfect 'profile shot' for an observer located anywhere within the Complex [of pyramids]. The line-like 'mouth' of this enigmatic object points directly toward the center of the city."

Richard Hoagland, Mars Investigation Team

"The Hoagland hypothesis is surely a possibility, but it sounds a bit 'Von Daniken' to me until we establish the city's existence beyond doubt."

Lambert Dolphin, Mars Investigation Team

dawn. And, a few moments afterward, the sun itself would have 'magically' appeared, rising directly out of the 'mouth' of the god-like figure [the face] in the Martian desert."

Hoagland calculates that such an alignment would occur about every million years. The last such alignment would have been half a million years ago. For this reason, Hoagland figures the Cydonia structures are at least half a million years old. Because the alignments are rare, Hoagland believes the Martians had the capability of predicting them. Thus they built the structures to prepare for the next occurrence.

Scientists Doubt Hoagland's Explanations

Anthropologist Jon Muller of Southern Illinois University believes Hoagland's calculations prove nothing. He says such alignments occur relatively often, and might be seen from any point in the solar system. There are so many alignments, in fact, that someone would have little trouble finding one to fit any theory.

Astronomer David Morrison is also skeptical of Hoagland's conclusions. He asks:

Do these [Martian monuments] represent one of the most important discoveries in human history? [Are they] destined to alter fundamentally our conceptions not only of Mars but of the origin and evolution of life and the nature of human consciousness? I doubt it.

Thus, despite Hoagland's claims, most scientists expressed little interest in exploring the Cydonia area further. Hoagland felt frustrated by this lack of interest. So he teamed with anthropologist Randolfo Rafael Pozos to arrange a conference with others who *were* interested.

Named after the Martian novel by science fiction author Ray Bradbury, the Martian Chronicles teleconference opened on December 5, 1983, in San Francisco. Some participants attended in person.

Others attended by hooking up to the conference with computer terminals from various locations across the country. Many of the participants were members of Stanford Research Institute International (SRI International), a private organization conducting research into controversial topics.

Chaired by Hoagland, the conferees debated the NASA photos and exchanged opinions. While some disagreed, most of the participants concluded that the Martian objects were not simply random products of nature but were purposefully built.

Physicist John Brandenberg stated that the Martian face showed clear evidence of construction. He said that it was balanced in appearance, with "two eyes, a nose, and a mouth. It appears to have an eye in one socket and to have cheek ornaments below the eyes. It is pleasing aesthetically. It looks like a king." He added that "other objects of non-natural appearance are found in the immediate area and

An illustration depicts a Viking landing on Mars. Participants at the Martian Chronicles conference believe that NASA should send additional missions to further explore the Cydonia area of Mars.

elsewhere." Based on these considerations, he said, "it is my judgment that this is an artificial object."

The City

Another conference participant was Lambert Dolphin, a senior research physicist with SRI International. He commented about the area Hoagland called the city. Dolphin said:

> The city is a rectangular set of cross hatches at the foot of the nearest mountain. I suppose the individual rooms are about 150 feet across, square, but also with obvious depth. This suggests to me a 'honeycomb,' as if a beehive had been opened from the top and a chunk carved out exposing cells at various depths.

He continued:

> Light seems to penetrate several levels, suggesting a building which was perhaps once enclosed in a dome or perhaps under a pyramid roof. The streets of the city more or less line up with the orientation of the face and also the edge of the mountain.

Conference participant Erol Torun, a map maker for the U.S. government, said that the Cydonia objects seemed to be the designs of humanlike beings. He also said that the objects were lined up in a way that proved the Martians knew something about astronomy and mathematics.

Following the conference, anthropologist and conference organizer Randolfo Pozos edited a book, *The Face on Mars*. In this book he reprinted transcripts of the Martian Chronicles discussions and summarized the participants' conclusion: The monuments of Cydonia were the ruins of an ancient, lost civilization.

Again, Hoagland's critics were quick to attack. "It is clear that the majority of the participants were not sufficiently cautious or scientific in their approach," says anthropologist Jon Muller. "Many of the participants were either directly or indirectly

This Viking photograph shows the pyramid-like structures that some people believe are the remains of an ancient city on Mars.

associated with SRI International, well known for its involvement in psychic and other borderline topics." In other words, the conference members believed the "ancient civilization" explanation was true even before the conference started. Therefore, there was no true scientific debate of the issues.

Muller says that some of the material in the Pozos book is unusually strange. It is strange enough, he says, "that one would almost suppose Pozos is doing this tongue-in-cheek." Muller adds that Pozos, in allowing questionable material into his book, "is certainly not well informed on either archaeology or the philosophy of science."

Objects Formed by Natural Processes?

Arden Albee, a California Institute of Technology geologist, says, "I don't know any scientists who believe that [the face is manufactured]. All around the world there are faces on mountains, so it just shows that the action of rocks falling and so on forms things that *look* like faces." He mentioned one earthly example, the Old Man of the Mountain in New Hampshire.

Of the pyramid, Harold Masursky, a specialist in the geology of other planets, says erosion probably caused the strange monuments. "In central Nevada," he says, "I have found a pyramid formed partly by stream erosion and faulting that's better than that. If you're going to say features like that are evidence for a past civilization, that's total nonsense."

Astronomer David Morrison also questions the interpretation of honeycomb patterns in the photographs as evidence of city streets or rooms. These, he says, are "an artifact of computer image processing." In other words, the honeycomb patterns in the photographs were not real objects. The patterns were created by the computer that was used to improve the pictures. The patterns then gave the false impression of a city on the Martian surface.

Quickly becoming the dominant voice in the

Geologists believe that the "face" on Mars, like the Old Man of the Mountain in New Hampshire (pictured), was probably formed by natural processes.

controversy, Richard Hoagland was not intimidated by the criticism. He formed the Independent Mars Investigation Team, headed by himself. IMIT's mission was to conduct ongoing research about the Martian monuments and to push for future photographic or human exploration of the Cydonia area. Dr. Mark M. Carlotto, an electrical engineer and image processing expert, was another member of the team. It was his job to analyze the photographs in a way that had never been done before.

Computer Enhanced Photographs Analyzed

Carlotto used new computer techniques to improve or enhance the two pictures of the face. One technique involved shape-from-shading. This technique produced a three-dimensional shape from the flat photographs. Another technique produced computer-printed pictures of the face. These pictures showed the object from many different viewpoints, as if a person were looking at the real object from different locations. This process could also show the object under different lighting conditions, as if the sun were shining on it from different angles. Thus Carlotto could print pictures of what the face would look like from above or from the ground, from any side of the structure.

These computer enhancements, Carlotto says, also enabled him to see details that were not apparent in the original photographs. These details include crossed lines above the eyes and fine structure in the mouth. Some people have referred to these mouth structures as teeth. Other details include regularly spaced stripes on the headpiece or helmet. Carlotto says that these features are real and not created by problems with the computer enhancement process. As a picture enhancement expert, he took special steps to make sure that no mistakes appeared.

Carlotto also discovered that the face's right side "is either incomplete or [worn away] and is not a mirror image of the left." Critics of the face see

"It is my belief that it [the D&M Pyramid] was not built, but carved out of a pre-existing mountain."

Richard Hoagland, Mars Investigation Team

"There happen to be some rock formations near the Face that look vaguely like pyramids, which is enough of a coincidence to turn the Face into a cult figure for some people."

Frank Drake, astronomer

Mystery continues to surround the planet Mars. Some people insist that the face and pyramid-like objects are evidence of an ancient civilization that once inhabited the planet.

this as evidence that the object is a naturally formed mesa. Supporters believe the right-side distortion was caused by meteorite impact, erosion, or abandonment of the construction project by its builders.

"The results of the 3-D analysis are plain," says Carlotto. "The impression of a recognizable facial structure is not [an optical illusion] as in New Hampshire's Old Man of the Mountain." The features are actually there, Carlotto says, and seem to show facial details regardless of the lighting conditions or viewing angles.

Astronomer David Morrison believes Carlotto's computer enhancement techniques are not very helpful. The three-dimensional computer model, he says, repeatedly uses artificial images in place of real information. "Such approaches may improve the images cosmetically," he says, "but they cannot generate information that was not present in the original."

NASA had hoped to continue its investigations

of Mars, including a 1993 mission to take additional photographs and make detailed maps of the Martian surface. Although there were no specific plans to photograph the area of the face, Hoagland's Mars Investigation Team hoped the new photos would help resolve questions about this mystery.

In fact, frustrated by NASA's lack of interest in the face, Hoagland turned to Congress to apply pressure on NASA. Representative Robert Roe, past chairman of the House Committee on Science, Space, and Technology, says:

> I've seen the studies and I've seen the photographs, and there do appear to be formations of a face and pyramids that do not appear to be of natural or normal existence. It looked like they had to be fashioned by some intelligent beings. . . . For this reason, I have asked NASA to provide assurance that the Mars Observer Mission will include this [set of objects] as one of its imaging objectives.

An unnamed NASA source said NASA had planned to take photographs of the Cydonia region, but that they intended to downplay "any specific angle of the so-called face." NASA officials believe the Cydonia objects are natural formations.

Answers May Have to Wait

In September 1993, the NASA Mars research effort underwent a major setback. The Mars Observer Satellite mysteriously lost contact with earth. Thus no new photographs will be available for research. Meanwhile Hoagland and his group are lobbying for astronauts to travel to Mars to explore the pyramids and face. Hoagland says, "The [Cydonia] data completely changes the status of going to Mars. There is a set of objects with wonderful and complex geometries on the Martian surface, and there seems to be a distinct possibility that someone made them. And the amazing thing is the prime advocates of SETI for the past twenty-five years—Carl Sagan, Frank

Photographs of the Martian surface could someday clear up the controversy surrounding the face.

Perhaps the answers will come someday when human exploration of Mars is possible.

Drake, Lou Friedman—don't seem to understand the implications."

One of those SETI advocates, astronomer Carl Sagan, voices the prevailing scientific opinion about going to Cydonia. "I'm not opposed to investigating," he says. "My view of the Face on Mars is my view on astrology. If someone can show that there is some validity to the claims, that's useful. But since the vast preponderance of the evidence is that it's nonsense, I don't think it's a good investment of resources." He believes that if an expedition goes to Mars solely to explore the Cydonia objects and then finds nothing but natural formations, it will hurt the prospects of ever going there again.

Sagan adds, "The idea of spending millions of dollars for an image which is intrinsically fuzzy just doesn't seem right."

The final answers about the Cydonia objects are still unknown. They remain unanswered until a human expedition explores the area. Mark Carlotto summarizes, saying that for those who argue over it, the face "is either completely natural or it is artificial. . . . There is no middle ground. . . . All will readily admit, however, that there is much to be learned."

Perhaps the final answer will come when someday humans walk among the pyramids of Cydonia.

"I have the perception that if the 'Martians' utilized existing landforms as architectural units fit for 'reshaping,' then this region of Mars located on the edge of this ancient 'coast' and amid abundant objects fit for 'reshaping' would have suited them admirably."

Richard Hoagland, Mars Investigation Team

"Hoagland never claims that any of this nonsense is true; he is forever interrupting himself to point out that these ideas are poorly founded, speculative, or incredible, before plunging back into his world of make-believe."

David Morrison, planetary astronomer, University of Hawaii

Five

Are Black Holes Doorways to Other Worlds?

Beyond the solar system is a vast universe filled with great mysteries. Many strange objects inhabit it, including billions of stars, beautiful glowing clouds of gas and dust, enormous galaxies, mysterious dark matter, and powerful sources of energy. Most of these objects generate light, heat, radiation, and other forms of electromagnetic energy. Thus scientists can detect and study them, using optical telescopes, radio telescopes, satellites, and other instruments. For the most part, scientists can measure temperatures and figure out what elements the space objects are made of. They can determine how far away the objects are and compute how fast they are receding from earth as the universe expands.

However, astronomers believe the universe contains other objects that no one can see or measure. They cannot be seen because they send out no light or radiation. That makes them invisible to telescopes and other instruments. Yet, scientists believe, certain of these objects are incredibly powerful. They bend the very space around them, altering the usual laws of physics. Like gigantic vacuum cleaners, they suck in other matter that comes too close. Matter falling into these objects disappears from the universe forever.

(Opposite page) Some scientists speculate that the entire universe will someday disappear, sucked into the unknown by black holes.

Even stranger, these bizarre objects may have powers once only dreamed of in science fiction stories. They may serve as pipelines or bridges to other parts of the universe or to other points in time. Or they may provide pathways to other universes. Most important, they may hold the keys to understanding the very nature and fate of the universe. These powerful, mysterious phenomena are called black holes.

Evidence of a Black Hole?

In 1992 NASA's Hubble telescope first photographed something scientists think is evidence of a black hole. Until then, scientists had no real evidence that they exist. Why, then, did they think black holes were real? For one thing, they are predicted by Einstein's theory of relativity, an important early-twentieth-century theory that shaped scientists' thinking about time and space. For another, scientists have detected strange behaviors in visible objects that are best explained by the influence of black holes. Proof of the existence of black holes will tell scientists a lot about the nature and fate of the universe.

In 1992 the Hubble Space Telescope (right) captured an image that scientists believe is evidence of a black hole.

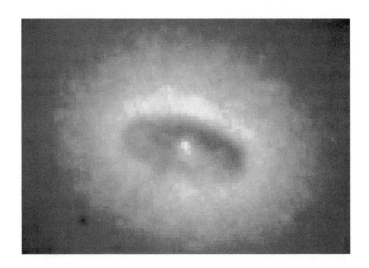

The Hubble Space Telescope captured this image of a giant disc of cold gas and dust. Scientists believe this disc is fueling a black hole.

A black hole is an extremely compact star with extraordinarily powerful gravity. The gravity is so strong that the star has collapsed, or caved in on itself. Dartmouth University space scientist Robert Jastrow says:

> The matter in the center of the star—all thousand trillion trillion tons of it—is squeezed from its original million-mile diameter down to a space the size of a house; then to the size of a golf ball; then to the size of a pin head; and then smaller, ever smaller.

All of this matter is crushed at the center into an infinitely small, unimaginably dense lump. When the star is squeezed down this much, the gravity is extremely powerful. It is so strong that not even light from the collapsed star escapes. (Scientists believe light is composed of particles called photons. The photons are influenced by gravity.) Jastrow says:

> Like a ball thrown upward from the surface of the earth, and then pulled back to the ground by gravity, every ray of light from the star is pulled back into its interior. Since no rays of light can get out, the star is invisible. It has become a black hole in space.

"Black holes eat stuff in an unforgiving, irreversible fashion. An object that falls into a black hole is forever removed from our universe. Since the object is no longer part of our universe, many of its properties are no longer detectable."

William Kaufmann III, astronomer, San Diego State University

"Black holes are the most grandiose energy sources in the universe. . . . It cannot be excluded that some day mankind will be able to utilize black holes as energy sources."

Igor Novikov, Russian astrophysics, University of Moscow

How does a star become a black hole? The story begins with the way stars live and die.

During the lives of most stars a constant battle is being waged. On one side is the expanding force of nuclear reaction, occurring deep in the star's interior. The energy from this force tends to push the star outward. On the other side is the contracting force of gravity. The pressure from this force tends to pull the star back in on itself. For many stars, including the sun, a long period of balance, or equilibrium, exists between the two forces. Stars enjoying such long periods of stability are known as main sequence stars. In the case of the sun, the main sequence is expected to last about 10 billion years. The sun is now approximately halfway through its main sequence.

A star in the main sequence, like most stars, is composed mostly of hydrogen and helium. It slowly consumes its supply of these fuels. Eventually the hydrogen begins to run out. As it does, the energy output decreases. Gravity starts winning the battle against the nuclear forces. The star's core slowly starts to collapse.

Red Giants

The collapse stops, however, when the star begins to burn helium instead of hydrogen. Continued burning of helium at ever higher temperatures forces the outer portion of the star outward again. Over tens of millions of years, the star swells to enormous size. It is now a very bright red, though relatively cool, star. Scientists call such stars red giants. (The earth's sun will enter this stage in about 5 billion years.)

The star lives as a red giant or an even larger supergiant for several million years. Then the helium runs out. Gravity again overcomes the outward pressure, pulling the star in on itself. This gravitational collapse results in a small white star called a white dwarf. This star is only about the size of the earth. Yet it is extremely compact. The entire star weighs about the same as the sun. Just a handful of matter from the star weighs five hundred tons. The white dwarf is not very bright. Its light is about one-thousandth of the light of our sun.

After a period of time, the last of the light and heat from the white dwarf radiates away. The star then becomes a black dwarf. The star is dead, a burned-out shell. Some scientists think this will be the ultimate fate of our sun.

Supernovas

A star that is about four to eight times larger than our sun, however, endures a different fate. When its fuel runs out, the star becomes unstable. Matter from the star's surrounding layers collapses into the core. As a result shock waves race outward, creating heavy elements, such as uranium, along the way. The star destroys itself in a violent explosion.

Physicist James Trefil says, "For a brief few days, the star can emit more energy than an entire galaxy. This event is a supernova—the most spectacular stellar cataclysm known."

The explosion hurls a fast-moving cloud of material into space. This material is rich in heavy elements. It later mixes with clouds of gas already in space to become a nursery for the birth of new stars. The only thing left of the star is its core.

The small arrow (left) points to a star that exploded to produce a supernova (right).

Sometimes a star's core continues to collapse following a supernova explosion. As it collapses the matter in this star becomes increasingly compressed, or dense. A mere thimbleful of this matter, for instance, would weigh about ten thousand tons. When the matter is this dense, the collapse stops. The weight of the star crushes atoms into pieces. Electrons and protons, now broken off from the atoms, begin to run into each other. Their collisions create subatomic particles called neutrons. Thus, the star becomes a neutron star.

Neutron stars are bizarre objects in their own right. Each has as much mass as several of our suns would have. Yet the star is only about ten to twelve miles in diameter. Strangely, the outer layer is solid, in the form of a crystalline crust. Underneath is a superfluid area similar to liquid helium. This material is extremely dense. A cube of this matter less than one-half cubic inch weighs 10 million tons!

Most neutron stars rotate rapidly, giving off flashes of light at their poles. Thus they are known as pulsars, or "pulsating stars."

How a Black Hole Is Born

When a neutron star is extremely massive, its gravity is so strong that its collapse continues. Its matter is crushed into an ever smaller space. Then gravity increases even more, and the matter is crushed further. Nothing can stop the process. Finally gravity becomes so powerful that it pulls light rays back into the star. As this happens the star, to an observer, would seem to fade away and then disappear. It has become a black hole, invisible and undetectable to the rest of the universe.

Even though unseen, however, the black hole is still there. It exerts extremely powerful gravity and pulls in anything that gets too close. Space scientist Robert Jastrow says:

Any ray of light or material that enters the black hole from the outside is also trapped; it can

"A black hole is . . . the extreme culmination pushed almost to absurdity of gravitational collapse."

Jean-Pierre Luminet, French astronomer

"Black hole radiation has shown us that gravitational collapse is not as final as we once thought."

Stephen Hawking, theoretical physicist, Cambridge University

Despite having a diameter of only ten to twelve miles, neutron stars have a mass several times the mass of the sun.

never get out again. The interior of the black hole is completely isolated from the outside world; it can swallow energy and matter, but it cannot send anything back. In effect, the material inside the black hole has been taken out of our universe. It has become a universe of its own.

To better understand how black holes work, it is helpful to picture the nature of gravity.

The Nature of Gravity and Space

All objects in space have gravity. Most commonly, we think of gravity as a force that attracts or pulls. However, most scientists think of gravity differently. The late science writer Isaac Asimov pictured space as a rubber sheet and gravity as dents or depressions in the sheet. He said, "Any heavy object resting on the sheet puts a dent in it. The heavier the object, the deeper the dent, and thus the stronger the gravity. If an object is kept heavy but made smaller, the weight is concentrated in a smaller area, and the dent gets deeper." The sheet of rubber (space), then, is warped, or bent, by the weight or mass of the object placed on it.

An illustration depicts space as an infinite elastic net. Every object in the universe that has mass sinks into this net and makes a depression. These depressions are what we experience as gravity.

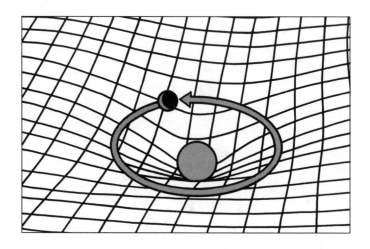

Thus a white dwarf makes a far deeper dent in the sheet than does the earth. A neutron star makes an even deeper depression. The deeper the dent, the more difficult it is to get out if you fall in. A black hole is so small and so heavy that it makes the deepest depression, one that is too deep to escape from.

Scientists say space is *curved* by gravity. This curvature is equivalent to the dents in the sheet of rubber. British mathematician John Taylor says, "Gravity is an intrinsic property of space. The way it is affected by heavy stars embedded in it is that it becomes curved. We notice this curvature when we travel past these stars, since we then move not in straight, but in curved lines around them."

Scientists have confirmed the curving, or warping, of space by measuring light rays from distant stars. They have found that the paths of light rays are often bent by the gravitational fields of other stars along the way.

The Inside of a Black Hole

No one really knows what the inside of a black hole is like. No light escapes from it to show what it is like. A satellite cannot go into it and report back. The satellite's signals, like the black hole's own light rays, would be pulled into the hole. Physicists and mathematicians, however, have made calculations to figure what the interior of a black hole might be like.

Near or at the edge of the black hole, they say, is a dangerous boundary. Anything passing this border will be captured by the black hole and be unable to turn around. This boundary is called the event horizon.

Astronomer William Kaufmann III of San Diego State University says, "It is literally a *horizon* in the geometry of space and time beyond which you cannot see any *events*. You have no way of knowing what happens inside an event horizon. It is a place disconnected from our space and time. It is not part of our universe."

No Way Out

British theoretical physicist Stephen Hawking adds, "The event horizon, the boundary of the region of space-time from which it is not possible to escape, acts rather like a one-way membrane around the black hole." In other words, it is something like a minnow trap. Minnows can swim in, but they cannot swim out. Hawking continues:

> Objects, such as unwary astronauts, can fall through the event horizon into the black hole, but nothing can ever get out of the black hole through the event horizon. Anything or anyone who falls through the event horizon will soon reach the region of infinite density and the end of time.

This region of infinite density is called the singularity. William Kaufmann explains how the singularity is formed:

> The strength of gravity and the curvature of space-time around the imploding [collapsing] star continue to grow, until the entire star is crushed down to a *single point*. At that point there is infinite pressure, infinite density, and most importantly infinite curvature of space-time.

Stephen Hawking says that falling into a black hole would bring you to a "region of infinite density and the end of time."

In other words, the singularity is the point where time and space cease to exist as humans know them. Time and space within the singularity are endless, and cannot be measured. Kaufmann continues, "This is where the star goes. Every atom and every particle in the star is completely crushed out of existence at this place. . . . This is the heart of the black hole." Scientists think the singularity exists in the shape of a ring.

Strange Effects

Scientists believe black holes do strange things to time and space. Outside the black hole the warping of space causes time itself to slow down, then to stop. To explain further, Kaufmann tells a story. In this story, he says, you are observing your friend fall toward a black hole. Your friend is holding a clock as he falls. "As your friend plunges toward a black hole, you see his clock running slower and slower. And at the moment of piercing the event horizon, the in-falling clock appears to stop completely as time forever stands still."

Your unfortunate friend, however, would not notice any slowdown in time. This is because *everything* slows down, including your friend's breathing, heartbeat, and thoughts. To him, time is passing by at a normal pace.

Scientists actively debate what would happen if someone fell into a black hole. Robert Jastrow says:

> The properties of black holes seem to suggest that he would be crushed by gravity. In actual fact he would be torn apart, because the part of his body closest to the center of the black hole would be pulled by a gravitational force stronger than any other part. . . . The astronaut would feel as though he were stretched on a rack. A few thousands of a second after entering the black hole, he would be dismembered. After a few more thousandths of a second, the individual atoms of his body would be broken into their

"A black hole is a cannibal, swallowing up everything that gets in its way. Once engorged by it there is no hope of escape; our own world is left behind forever on passing into its event horizon."

John Taylor, mathematics professor, University of London

"If an astronaut falls into a black hole, he will be returned to the rest of the universe in the form of radiation. Thus, in a sense, the astronaut will be recycled. However, it would be a poor sort of immortality because any personal concept of time would come to an end as he is torn apart inside a black hole."

Stephen Hawking, theoretical physicist, Cambridge University

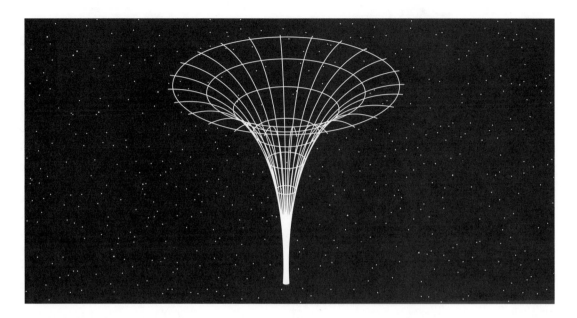

separate neutrons, protons, and electrons. Finally the elementary particles themselves would be torn into fragments whose nature is not yet known to physicists.

Some scientists suggest there may be ways to avoid this fate. William Kaufmann believes that an astronaut would only be doomed if he or she approached the ring-shaped singularity from its edge. If the approach was from any other angle, the astronaut would pass through the ring unharmed.

Strange Possibilities

Passing through the ring, however, invites some very strange possibilities. One of the possibilities, Kaufmann says, is that "you do not simply come out the other side. Instead, you enter negative space, where gravity is repulsive. . . . You enter an anti-gravity universe, a place where gravity pushes things up rather than pulls things down!" In this strange universe objects would fall away from a planet instead of toward it. If you fell off a cliff, you would fall up instead of down.

Because of its extreme density, the gravity well of a black hole is very deep. While scientists agree that it would be impossible to escape, there is considerable debate about what would happen if someone fell into a black hole.

Other theorists suggest that black holes may serve as doorways to other universes or to other points in time. Scientists call such bridges wormholes. Scientists have no physical evidence that wormholes exist, but mathematical calculations show that they are a possibility. And mathematical calculations have predicted many things about the universe that have proven to be true. For example, long before Pluto was discovered, calculations predicted that a ninth planet circles the sun.

Remember the comparison of space to a rubber sheet. Imagine that the sheet folds under itself, forming a shape somewhat like the letter *U* placed sideways. The top portion of the shape represents our universe, or our time. The bottom part represents another universe, or another time in our own universe.

On the top part of the rubber sheet is a black hole, making a deep dent. Imagine that the dent reaches down to the lower part of the rubber sheet and opens out on the other side. This connecting funnel is the wormhole. It serves as a path from our universe to another, or from our period of time to another. Scientists call the wormhole exit a white hole.

Wormholes: Doorways to Another Universe?

According to theory objects or people might be able to pass through a wormhole to other locations in time and space. The angle of approach would determine which of the possibilities would occur. For example, a steep angle might result in a trip to an antigravity universe, such as the one described by Kaufmann. A shallow angle might result in a trip to another universe like ours. Other angles might result in a trip to yet another universe, or a trip to another time or place in our own universe. Kaufmann says:

> Perhaps a rotating black hole connects our universe with itself in a multitude of places. But remember, these would be different places in space and/or time. In other words, by emerging

into one of these 'other universes,' you might actually be reentering our own universe in the same place but at a different time. This is a time machine!

Such a thing is possible, according to Kaufmann, if the pilot guides the spacecraft carefully. "You could re-emerge into our universe a billion years ago and visit the earth before the age of dinosaurs," he says. "Or you could emerge a billion years in the future and meet the creatures that eventually evolved from the lower life-forms that today we call human beings." Furthermore, Kaufmann says:

> If you could use a rotating wormhole as a time machine and come back a billion years ago, you could also certainly arrange to come back to the earth an hour before you left. You could meet yourself and tell yourself what a fine trip you had! And then both of you could get on the rocket ship and take the trip again! And again! And again!

Stephen Hawking, however, is doubtful that black holes will ever provide shortcuts for space or time travelers. There are problems, he says, in getting to your intended destination. "If one could pass

An illustration of a black hole. Some people believe a black hole could serve as a doorway to another location in time and space.

through a black hole, one might reemerge anywhere in the universe. Quite how you choose your destination is not clear. You might set out for a holiday in [the] Virgo [constellation] and end up in the Crab Nebula." Besides, he says, humans probably will never get far enough into the black hole to enter the wormhole:

> I'm sorry to disappoint prospective galactic tourists, but this scenario doesn't work. If you jump into a black hole, you will get torn apart and crushed out of existence. However, there is a sense in which the particles that make up your body do carry on to another universe. I don't know if it would be much consolation to someone being made into spaghetti in a black hole to know that his particles might survive.

John Taylor agrees with Hawking. "It may not be till the atoms of the astronaut and his spaceship have been compressed about a billion billion times at the black hole singularity that these gates open and allow the matter to fly freely through superspace," he says. "By then even the atoms themselves, let alone the astronaut and his spaceship, will have completely lost their identity. So the possibility of using superspace to achieve instantaneous journeys or even time travel seems rather remote."

Nonetheless, Taylor maintains that black holes will be tempting to those bold enough to approach them in the future. He says:

> The particular way in which time is slowed down on the edge of a black hole is especially interesting, so leading the way to a possible time machine. It would [give] near immortality to those venturesome enough to travel close to the event horizon, always staying on the safe side. Not that they would feel any older themselves; their life span would only have been lengthened with respect to their sit-at-home friends and relations.

For the last two decades scientists have devoted

considerable time to the search for black holes. The search has been difficult, however, because of black hole invisibility. Scientists have had to find a way to detect them indirectly.

"The most obvious thing to look for," says John Taylor, "is the possible effects black holes would have on surrounding stars. If there were any black holes, then stars would move in an incomprehensible fashion, jostled this way and that by the nearby marauders."

Such jostling might best be detected in a binary star system. That is, a system of two stars, one revolving around the other in much the way the planets of our solar system revolve around the sun. Scientists think they would see the strongest jostling in a binary star system made up of one normal star and one black hole. They would see only the normal star, but it would probably have an unusual orbit,

An artist's conception of a black hole sucking in the matter of a star. Some astronomers believe that supermassive black holes exist, slowly drawing in millions, even billions, of nearby stars.

caused by the gravity of the black hole. Light passing the area of the unseen companion might bend in a strange way. Matter from the normal star might be seen getting sucked away from the star and into the area of the black hole. A powerful flow of gas from the envelope of the normal star would pass into the gravitational field of the black hole companion and form a spiral shape around the black hole. This spiral is called an accretion disk. It would give off X rays, lose energy, and finally fall into the black hole.

In searching the sky for this kind of system, scientists cannot see the black hole itself. But they can detect the X rays produced by the accretion disk. Such X-ray sources have in fact been found. One strong candidate is located in the constellation Cygnus. Astronomers have detected X rays coming from Cygnus X-1, a large star that is apparently revolving around an unseen companion. Scientists believe the companion is a black hole.

Astronomers suspect there is another kind of black hole in the universe: the giant, or supermassive, black hole. Such black holes would be huge, about the size of our solar system, residing in the cores of galaxies, slowly drawing in millions, even billions, of nearby stars.

Quasars

Indeed scientists have detected distant objects radiating incredible amounts of energy into space. Scientists call them quasars for "quasi-stellar" or starlike energy sources. These objects, which may be the side effects of black holes, exist at the farthest reaches of the observable universe. Jastrow says that the possible existence of giant black holes explains two aspects of quasars. "First," he says, "it accounts for the prodigious amount of energy coming from a quasar. Second, since black holes are exceedingly compact objects, it accounts for the fact that the energy comes from a very small region in space."

"There are other possible ways out of the terrible dilemma at the center of a black hole. It may be that the larger wormholes reach out to take the pulverized matter from the singularity and let it bubble up somewhere else in our universe. It is as if there were a tube from the central point that leads out of the event horizon."

John Taylor, mathematics professor, University of London

"The black hole may be bottomless, or it may be a tunnel leading to another universe. There is no way for science to know, and all speculation about wormholes and white holes is exactly that: speculation."

Neil McAleer, space science writer

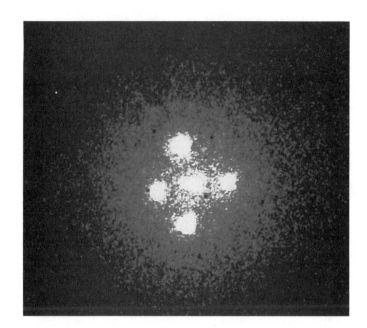

Quasars, which have been detected in the farthest reaches of the universe, radiate massive amounts of energy into space.

With the help of the orbiting Hubble Space Telescope, scientists have found other evidence of possible supermassive black holes. High-resolution photographs reveal tremendous clusters of stars at the center of the nearby galaxy M32. The densely grouped stars there are considered to be the gravitational signature of a massive black hole. And extremely hot gas concentrations are detected at the center of galaxy NGC 4261, about forty-five light-years away. Scientists say the gases are made by a gigantic accretion disk formed around a supermassive black hole.

About thirty light-years away, in the constellation Sextans, is NGC 3115. The stars in this galaxy are moving extremely fast. Scientists conclude that only a black hole with the mass of a billion suns could keep these stars from flying completely out of the galaxy.

There are other black hole candidates, and scientists are actively searching for more. The evidence is stronger than ever that black holes, once only the product of mathematical calculations, are very real.

Black Holes and the Fate of the Universe

What is the fate of a universe full of black holes? Will there eventually be so many black holes that they will pose a hazard to space travelers? Will they continue to gobble matter until none is left? Will the entire universe end as a gigantic black hole, with all that there is—planets, stars, galaxies, other black holes, everything—compressed into a final singularity?

Yes, says John Taylor. "As far as we understand black holes, and more generally the nature of space and time," he says, "such a fate is inevitable. Billions of years will pass, but the black holes will get us in the end." The only choice, he says, is to develop the ability to escape through a black hole to another universe, where things might be "less dangerous."

Stephen Hawking disagrees. He says the threat of black holes may be exaggerated. He believes that

Will black holes continue to gobble matter until the entire universe is gone? The fate of the universe is a mystery that continues to intrigue us.

black holes do not actually keep everything they take in. He contends that they leak a small amount of radiation. While this amount of leakage is small, it is enough to eventually allow the black hole to spread out. Over time the black hole would stop drawing in matter. Eventually it would break apart, ending the threat.

The Possible Destinies

Still, Hawking agrees with Taylor about the ultimate fate of the universe. He says, "The universe has two possible destinies. It may continue to expand forever, or it may recollapse and come to an end at the big crunch. I predict the universe in time will come to an end at the big crunch."

"The other alternative is a re-expansion after compression," says John Taylor. He continues:

> In that case the whole story would start all over again—expansion, slowing down, maximum expansion point finally being reached, contraction, ultimately with destruction of the significant features we see around us. Such a recycling would continue indefinitely, and may have already done so in the past.

In either case it is doubtful that humanity will survive the effects of the big crunch. Taylor says, "Even the fundamental particles of which we are composed—the electron, proton, and neutron—will very likely be decimated [destroyed] as well, though long after ourselves. There appears to be no escape for us in either situation."

Taylor believes humans must take heart in the possibility that, following the big crunch, a new universe will explode into existence. In this new universe, new stars will shine, and new worlds will form. In some of these worlds, new life-forms will appear, possibly beings similar to humans. In that sense intelligent life-forms may survive through a multitude of new universes to eternally ponder the same questions we ask today.

"Our ultimate fate is collapse. Beyond that is the still unknowable part of infinity."

John Taylor, mathematics professor, University of London

"Whether we live in an open or closed universe is currently unknown. The bottom line clearly suggests an evolutionary universe, but its ultimate destiny remains concealed."

Eric Chaisson, astrophysicist, Harvard University

Conclusion

The Universe Calls

We are fortunate to live in a great age of discovery. New satellites orbit the earth. Bigger and better telescopes study the stars. All of these things promise to reveal many more secrets of the universe. Many of the answers we seek may come in our lifetimes. But will we ever determine the ultimate truth about the universe?

The great scientist Albert Einstein spent most of his life trying to develop what he called a grand unified theory. This theory would unite all the sciences. It would explain in simple terms the nature of the entire universe, from the large to the small. Einstein failed in this quest.

Scientists Search for Clues

Today scientists study new forms of science, including quantum physics, that grew out of and beyond Einstein's revolutionary ideas. Quantum physics is the study of subatomic particles, the tiniest particles of all energy and matter. Scientists hope that what they learn about these tiny particles will give more clues leading to a grand unified theory.

Stephen Hawking is one who has taken part in this search. He says:

If we do discover a complete theory, it should in time be understandable in broad principle by

(Opposite page) Scientists will continue to search the skies for clues to unravel the mysteries of the universe.

everyone, not just a few scientists. Then we shall all, philosophers, scientists, and just ordinary people, be able to take part in the discussion of the question of why it is that we and the universe exist. If we find the answer to that, it would be the ultimate triumph of human reason—for then we would know the mind of God.

The search for answers goes on.

Glossary

accretion disk   a gathering of matter in a spiral pattern around a compact object such as a newly forming star or a black hole.

amino acid   an organic compound. Certain kinds of amino acids are the building blocks of proteins.

astronomy   the observation and study of the heavens, including stars, planets, galaxies, and other natural objects in space.

astrophysics   the study of the physical makeup and characteristics of objects in space.

atmosphere   the gases surrounding a planet, moon, or star.

atom   the smallest particle of an element (such as hydrogen or helium).

bandwidth   a group of signals that are broadcast or received within a limited frequency range. For example, radio stations are on a radio bandwidth; television stations are on a television bandwidth. The search for signals from other worlds is conducted within the water hole bandwidth.

big bang   the ancient explosion that many scientists believe launched the expansion of the universe. Scientists believe it happened about 15 to 20 billion years ago.

binary star   a double star system in which two stars orbit each other, influenced by the gravity of both stars.

biochemistry   the branch of chemistry that studies the chemical processes of living things.

biological evolution change in plants and animals over a long period of time.

black dwarf a burned-out star that no longer radiates energy.

black hole a burned-out star that has collapsed. Its extremely dense mass and powerful gravity keep light and radiation from escaping. It is therefore invisible to the rest of the universe.

cell a basic unit of living matter.

chemical evolution the development of simple life, over a long period of time, from nonliving material.

cosmic background radiation a microwave radiation left over from the big bang fireball. This radiation comes to earth from all directions in the universe.

cosmology the study of the origin and nature of large structures in the universe.

dark matter invisible matter thought to exist in and around galaxies. Scientists believe dark matter makes up about 90 to 99 percent of the total matter in the universe. Dark matter may serve as a sort of glue holding the visible universe together.

density the amount of compactness or solidity of matter. Rock, for instance, has greater density than water.

directed panspermia the planned "seeding" of life on earth by intelligent beings from other worlds.

Doppler effect a wavelength shift that causes sound, such as a police siren, to rise in pitch as it approaches. The same effect causes the sound to decrease in pitch as it departs. The Doppler effect also applies to shifts in wavelengths of light. Approaching light causes a shortening of the light wavelengths, recorded by machines as a shift to the blue end of the spectrum. Departing light causes a lengthening of the light wavelengths, recorded as a red shift. The red shift of most objects in the universe tells scientists that the universe is moving away, or expanding.

Drake equation a formula developed by astronomer Frank Drake. The formula estimates the number of civilizations in the galaxy capable of communicating with one another.

electromagnetic spectrum the full range of radiation and energy in the universe. This includes radio waves, light waves, microwaves, ultraviolet rays, and X rays.

electron a basic unit of matter that makes up part of an atom. It has a negative electrical charge.

event horizon a border around the center of a black hole. Anything falling through this boundary is thought to be forever trapped inside the black hole.

evolution change over a long period of time.

extraterrestrial outside the earth or its atmosphere. In astronomy this refers to objects in space, such as stars, planets, and so on. In SETI it refers to beings from another world.

frequency the number of cycles or vibrations of a wave motion per second.

galaxy a large, rotating grouping of stars, gas, and dust, held together by mutual gravity. Most galaxies are millions of light-years in diameter.

grand unified theory a theory that explains the electromagnetic and subatomic aspects of the universe in one statement. No one has yet developed this theory.

gravitation the basic force of attraction that matter exerts on other matter.

helium a light, colorless gas that is a common part of all stars. A helium atom has two protons and two electrons.

hydrogen   a light, colorless gas that is the most common chemical element in the universe. The hydrogen atom has one proton and one electron. Most stars are 75 percent hydrogen.

infinite   endless, limitless, boundless, and without measure.

infinity   eternal endlessness; unlimited.

interstellar   the space between stars.

light-year   the distance light travels in a year, nearly 6 trillion miles.

main sequence   the stable period of a star's life, when it is consuming mostly hydrogen.

mass   the amount of matter packed into an object.

matter   the material the universe is made up of. Matter occupies space, exists in time, and is detectable to instruments and human senses.

microwaves   radio waves with wavelengths between one millimeter and a few centimeters.

Milky Way   the name for the galaxy in which the earth exists. It is named for its appearance in the nighttime sky. On a clear, moonless night, it appears as a bright, "milky" path in the sky. The "path" is actually one spiral arm of the galaxy, where millions of stars are relatively close together.

molecule   the smallest unit of an element (such as iron) or compound (such as water). It is composed of atoms.

multichannel spectrum analyzer (*also* multichannel spectrometer)   a computer-aided receiver that analyzes signals at different frequencies at the same time.

NASA   the National Aeronautics and Space Administration. NASA manages the U.S. space program. This program includes many projects, such as shuttle launches, satellite exploration of the planets, space station design, and the search for life on other worlds.

nebula a loose cloud of gas and dust in the space between stars. Sometimes these clouds reflect the light of distant stars. Thus many nebulae appear in photographs as beautiful, brightly colored objects.

neutron a basic unit of matter with no electrical charge. It is usually found in the nucleus, or center, of an atom.

neutron star the small, densely packed core left over from a supernova. It is composed mostly of neutrons.

panspermia the "seeding" of life on the early earth or other planets by spores from other worlds.

photosynthesis a process by which plants convert the sun's energy to oxygen and nutrients.

planet a world circling a star that reflects rather than gives off light.

plasma an energized gas.

proton a basic unit of matter that has a positive electrical charge. It is normally found in the nucleus, or center, of an atom.

pulsar a rapidly rotating neutron star that sends out pulses of light, X rays, and gamma rays.

quasar an active starlike nucleus of a distant galaxy. These are the most distant known objects in the universe. A quasar is usually brighter than hundreds of normal galaxies. The energy source may actually be a large black hole at the quasar's center.

radio astronomy the observation and study of radio waves sent out by objects in space.

radio telescope a dish antenna system that receives and sends radio signals. Scientists use radio telescopes to detect radio waves put out by distant objects in space.

radio wave an electromagnetic wave or signal that can be detected by radio telescopes and other instruments.

red giant   a star that has used up most of its hydrogen fuel. It has left the main sequence and swollen as much as a hundred times its original size. Its outer surface is red and relatively cool when compared to yellow or white stars.

red shift   the reddening of light sent out by objects that are moving away. The greater the red shift, the faster the object is moving away. The red shift is caused by the Doppler effect. Measuring it tells how far away an object is and how fast it is moving.

SETI   the Search for ExtraTerrestrial Intelligence. Most SETI projects use radio telescopes to search the skies for signals from other worlds. The detection of such signals would prove that humans are not the only intelligent forms of life in the universe.

solar system   the sun, its planets, and all other objects orbiting the sun.

space-time   the flow of time and the movement of objects through space. According to Einstein, both space and time are aspects of the same thing. One cannot exist without the other.

star   a round body in space such as the sun that radiates energy, including light. This energy is generated by nuclear processes in the star's interior.

steady state theory   a cosmological theory which states that the universe has no beginning and no end and appears to change little over time. As the universe expands, new matter is created to replace the matter that has moved away.

supernova   a large exploding star. It briefly outshines an entire galaxy of stars, then fades in a matter of weeks or months. The explosion hurls material from the star into space, seeding the cosmos with material for future generations of stars. Left behind is a core that will become a neutron star or a black hole.

universe   all that exists or is believed to exist. This includes large things, such as interstellar gas and dust, planets, stars, quasars, pulsars, black holes, galaxies, and the space in between. It also includes small things, such as people, animals, plants, microbes, molecules, and atoms.

water hole   a band of radio frequencies in an area of the electromagnetic spectrum that is relatively free of radio interference. It is between the frequencies of hydrogen and hydroxyl, the components of water. Scientists consider it is a likely range in which to find signals from extraterrestrial beings.

wavelength   the distance between two successive wave crests or troughs in a wave of light or other forms of electromagnetic energy.

white dwarf   a small, white-hot star left over from the collapse of a star like the sun. It is about the size of the earth.

wormhole   the connecting funnel to a black hole. According to theory, it would serve as a path from our universe to another, or from our period of time to another.

X rays   particles with energies greater than those of ultraviolet light and less than those of gamma rays.

For Further Exploration

Isaac Asimov, *How Was the Universe Born?* New York: Dell, 1991.

———, *Quasars, Pulsars, and Black Holes.* New York: Dell, 1990.

Franklyn M. Branley, *Is There Life in Outer Space?* New York: Crowell Junior Books, 1984.

Mark J. Carlotto, *The Martian Enigmas: A Closer Look.* Berkeley, CA: North Atlantic Books, 1991.

Frank Drake and Dava Sobel, *Is Anyone Out There? The Scientific Search for Extraterrestrial Intelligence.* New York: Delacorte, 1992.

Dwight Dwiggins, *Hello? Who's Out There?* New York: Dodd, Mead, 1987.

Robert Jastrow, *Journey to the Stars.* New York: Bantam, 1989.

Neil McAleer, *The Cosmic Mind-Boggling Book.* New York: Warner, 1982.

Patrick Moore, *Astronomy for the Beginner.* New York: Cambridge University Press, 1992.

———, *The Universe for the Beginner.* New York: Cambridge University Press, 1992.

Richard Michael Rasmussen, *Extraterrestrial Life.* San Diego: Lucent, 1991.

Carl Sagan, *Other Worlds.* New York: Bantam, 1975.

Works Consulted

Chapter 1
How Did the Universe Begin?

Eric Chaisson, *Cosmic Dawn: The Origins of Matter and Life.* Boston: Atlantic Monthly Press/Little, Brown, 1981.

William K. Hartmann and Ron Miller, *The History of Earth: An Illustrated Chronicle of an Evolving Planet.* New York: Workman, 1991.

Eric J. Lerner, *The Big Bang Never Happened: A Startling Refutation of the Dominant Theory of the Origin of the Universe.* New York: Vintage, 1992.

Dennis Overbye, "Cosmologies in Conflict," *Omni,* October 1992.

———, *Lonely Hearts of the Cosmos.* New York: HarperCollins, 1991.

Ivars Peterson, "State of the Universe: If Not with a Big Bang, Then What?" *Science News,* April 13, 1991.

Evry Schatzman, *Our Expanding Universe.* New York: McGraw-Hill, 1992.

Victor J. Stenger, "Is the Big Bang a Bust?" *Skeptical Inquirer,* Summer 1992.

Richard Talcott, "COBE's Big Bang," *Astronomy,* August 1992.

James S. Trefil, *The Moment of Creation: Big Bang Physics.* New York: Scribner's, 1983.

Chapter 2
How Did Life Form on Earth?

Svante Arrhenius, "The Propagation of Life in Space," 1903 article reprinted in *The Quest for Extraterrestrials*, Donald Goldsmith, ed. Mill Valley, CA: University Science Books, 1980.

Francis Crick and Leslie Orgel, "Directed Panspermia," 1972 article reprinted in *The Quest for Extraterrestrials,* Donald Goldsmith, ed. Mill Valley, CA: University Science Books, 1980.

Richard E. Dickerson, "Chemical Evolution and the Origin of Life," 1978 article reprinted in *The Quest for Extraterrestrials,* Donald Goldsmith, ed. Mill Valley, CA: University Science Books, 1980.

Robert M. Hazen and James Trefil, *Science Matters: Achieving Scientific Literacy.* New York: Doubleday, 1991.

Fred Hoyle, *The Intelligent Universe: A New View of Creation and Evolution.* New York: Holt, Rinehart, and Winston, 1984.

Fred Hoyle and N.C. Wickramasinghe, *Evolution from Space: A Theory of Cosmic Creationism.* New York: Touchstone Books, 1981.

————, *Lifecloud: The Origin of Life in the Universe.* New York: Harper & Row, 1978.

Joshua Lederberg, "Exobiology: Approaches to Life Beyond the Earth," 1960 article reprinted in *The Quest for Extraterrestrials,* Donald Goldsmith, ed. Mill Valley, CA: University Science Books, 1980.

John Reader, *The Rise of Life: The First 3.5 Billion Years.* New York: Knopf, 1986.

P.H.A. Sneath, *Planets and Life.* New York: Funk and Wagnalls, 1970.

Chapter 3
Does Life Exist Beyond the Solar System?

Joseph A. Angelo Jr., *The Extraterrestrial Encyclopedia: Our Search for Life in Outer Space.* New York: Facts on File, 1985.

John Billingham, ed., *Life in the Universe.* Cambridge, MA: MIT Press, 1981.

Ben Bova and Byron Preiss, eds., *First Contact: The Search for Extraterrestrial Intelligence.* New York: NAL, 1990.

James L. Christian, ed., *Extraterrestrial Intelligence: The First Encounter.* Buffalo, NY: Prometheus Books, 1976.

Arthur C. Clarke, "Why Is It Important?" *Life,* September 1992.

Frank Drake and Dava Sobel, *Is Anyone Out There? The Scientific Search for Extraterrestrial Intelligence.* New York: Delacorte, 1992.

Thomas R. McDonough, "Searching for Extraterrestrial Intelligence," *Skeptical Inquirer,* Spring 1991.

Edward Regis, ed., *Extraterrestrials: Science and Alien Intelligence.* New York: Cambridge University Press, 1985.

David Leigh Rodgers, *World Alone.* Roslyn Heights, NY: Libra, 1974.

Robert T. Rood and James S. Trefil, *Are We Alone? The Possibility of Extraterrestrial Civilizations.* New York: Scribner's, 1981.

Seth Shostak, "Listening for Life," *Astronomy,* October 1992.

Dava Sobel, "Is Anybody Out There?" *Life,* September 1992.

Frank White, *The SETI Factor.* New York: Walker, 1990.

Chapter 4
Is Mars the Site of Ancient Cities?

Joseph Baneth Allen, "That Face on Mars Again," *Final Frontier,* July/August 1991.

Mark J. Carlotto, *The Martian Enigmas: A Closer Look.* Berkeley, CA: North Atlantic Books, 1991.

Richard Grossinger, *The Night Sky.* Los Angeles: Jeremy Tarcher, 1988.

Richard C. Hoagland, *The Monuments of Mars: A City on the Edge of Forever.* Revised, expanded edition. Berkeley, CA: North Atlantic Books, 1992.

Jay Lebsch, "Mars Mission Draws Space Community into New Orbits," *UFO,* November/December 1988.

David Morrison, "Seeing Faces on Mars," *Skeptical Inquirer,* Fall 1988.

Jon Muller, "Wishful Seeing," *Skeptical Inquirer,* Fall 1987.

Randolfo Rafael Pozos, ed., *The Face on Mars: Evidence of a Lost Civilization?* Chicago: Chicago Review Press, 1987.

Chapter 5
Are Black Holes Doorways to Other Worlds?

Stephen Hawking, *A Brief History of Time: From the Big Bang to Black Holes.* New York: Bantam, 1988.

Stephen Hawking, ed., with Gene Stone, *Stephen Hawking's a Brief History of Time: A Reader's Companion.* New York: Bantam, 1992.

Robert Jastrow, *Journey to the Stars.* New York: Bantam, 1989.

William J. Kaufmann III, *Black Holes and Warped Space-Time.* San Francisco: W.H. Freeman & Co., 1979.

Jean-Paul Luminet, *Black Holes.* Cambridge: Cambridge University Press, 1992.

Igor Novikov, *Black Holes and the Universe.* Cambridge: Cambridge University Press, 1990.

John G. Taylor, *Black Holes: The End of the Universe?* New York: Random House, 1973.

Additional Works Consulted

Isaac Asimov, *The Collapsing Universe.* New York: Walker, 1977.

———, *Extraterrestrial Civilizations.* New York: Crown, 1979.

———, "Science and the Mountain Peak," *Skeptical Inquirer*, Winter 1980-81.

———, *The Subatomic Monster.* New York: NAL/Mentor Books, 1986.

David Attenborough, *Life on Earth.* Boston: Little, Brown, 1979.

"A Black Hole for M32," *Astronomy,* July 1992.

Zen Faulkes, "Getting Smart About Getting Smarts: Evolutionary Biology and Extraterrestrial Intelligence," *Skeptical Inquirer*, Spring 1991.

Gerald Feinberg and Robert Shapiro, *Life Beyond Earth: The Intelligent Earthling's Guide to Life in the Universe.* New York: William Morrow, 1980.

Timothy Ferris, *The Red Limit: The Search for the Edge of the Universe.* New York: Quill, 1983.

Tim Folger, "In the Black," *Discover*, January 1993.

Herbert Friedman, *The Astronomer's Universe: Stars, Galaxies, and Cosmos.* New York: Ballantine, 1990.

Larry Gels and Fabrice Florin, eds., *Worlds Beyond: The Everlasting Frontier.* Berkeley, CA: And/Or Press, 1978.

Frederic Golden, *Quasars, Pulsars, and Black Holes.* New York: Scribner's, 1976.

John Gribbon, *In Search of the Big Bang.* New York: Bantam, 1986.

———, *Timewarps.* New York: Delta Books/Dell, 1980.

Nick Herbert, *Faster than Light: Superluminal Loopholes in Physics.* New York: NAL, 1988.

"Hubble Views Possible Black Hole," *Astronomy*, March 1993.

Michio Kaku and Jennifer Trainer, *Beyond Einstein: The Cosmic Quest for the Theory of the Universe.* New York: Bantam, 1987.

Rudolf Kippenhahn, *100 Billion Suns: The Birth, Life, and Death of the Stars.* New York: Basic Books, 1983.

Neil McAleer, *The Omni Almanac: A Complete Guide to the Space Age.* New York: Omni Books/World Almanac, 1987.

John Macvey, *Time Travel: A Guide to Journeys Beyond the Fourth Dimension.* Chelsea, MI: Scarborough House, 1990.

Jerry Pournelle, ed., *Black Holes.* New York: Fawcett Crest, 1978.

Michael Riordan and David N. Schramm, *The Shadows of Creation: Dark Matter and the Structure of the Universe.* New York: W.H. Freeman, 1991.

Carl Sagan, *Cosmos.* New York: Random House, 1980.

Walter Sullivan, *Black Holes: The Edge of Space, the Edge of Time.* New York: Anchor Books/Doubleday, 1979.

James Trefil, *The Dark Side of the Universe: A Scientist Explores the Mysteries of the Cosmos.* New York: Anchor, 1989.

John M. Williams, "The Star Hunters," *Final Frontier*, November/December 1991.

Fred Alan Wolf, *Parallel Universes: The Search for Other Worlds.* New York: Touchstone, 1990.

Index

About the Author

Richard Michael Rasmussen is a professional writer living in La Mesa, California. Rasmussen holds an A.A. degree from Grossmont College and a certificate in technical writing from San Diego State University. He is the author of six other books, including two published by Lucent Books—*The UFO Challenge* and *Extraterrestrial Life*. Rasmussen is a member of the Planetary Society, a space-advocacy group, and Southern California Skeptics, a group promoting the scientific way of investigating things. As a member of the national Society of Children's Book Writers and the San Diego Writers and Editors Guild, he gives talks to elementary and junior high schools in hope of inspiring young people to become professional writers.

Picture Credits